KU-480-139

Your Book of Canals

The Your Book *Series*

Acting
Aeromodelling
Anglo-Saxon England
Animal Drawing
Aquaria
Astronomy
Badminton
Ballet
Brasses
Breadmaking
Bridges
Butterflies and Moths
Cake Making and Decorating
Camping
Canals
The Way a Car Works
Card Games
Card Tricks
Chess
Coin Collecting
Computers
Confirmation
Contract Bridge
Corn Dollies
Mediaeval and Tudor Costume
Nineteenth Century Costume
Cricket
Dinghy Sailing
The Earth
Embroidery
Engines and Turbines
Fencing
Film-making
Fishes
Flower Arranging
Flower Making
Golf
Gymnastics
Hovercraft
The Human Body
Judo
Kites
Knitted Toys
Knots
Landscape Drawing
Light
Magic
Men in Space
Mental Magic
Modelling
Money
Music
Painting
Paper Folding
Parliament
Party Games
Patchwork
Patience
Pet Keeping
Photography
Photographing Wild Life
Keeping Ponies
Prehistoric Britain
Puppetry
Racing and Sports Cars
The Recorder
Roman Britain
Sea Fishing
The Seashore
Secret Writing
Self-Defence
Shell Collecting
Skating
Soccer
Sound
Space Travel
Squash
Stamps
Surnames
Swimming
Survival Swimming and Life Saving
Swimming Games and Activities
Table Tennis
Table Tricks
Television
Tennis
Traction Engines
Veteran and Edwardian Cars
Watching Wild Life
The Weather
Woodwork

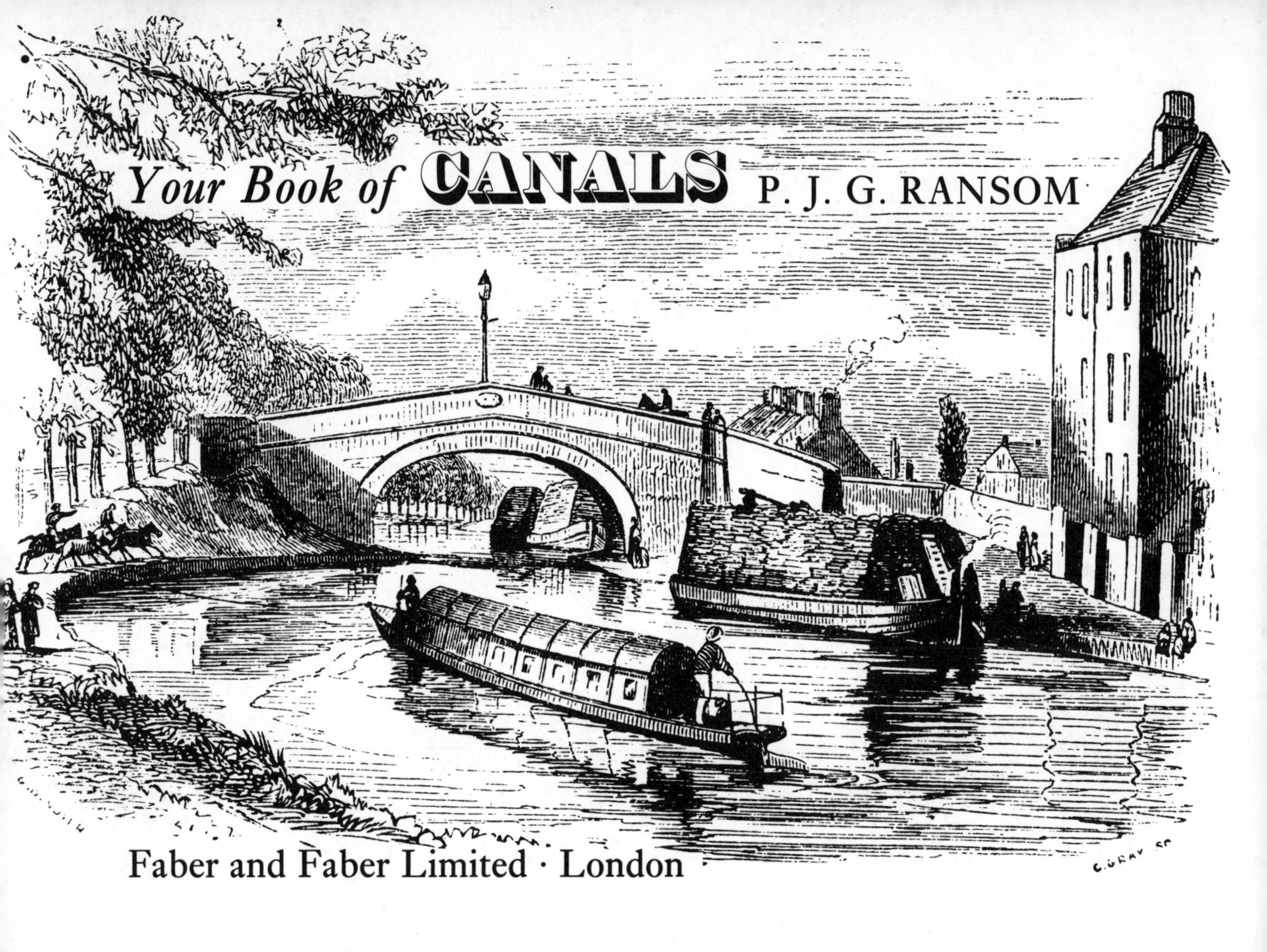

Your Book of CANALS

P. J. G. RANSOM

Faber and Faber Limited · London

First published in 1977
by Faber and Faber Limited
3 Queen Square London WC1
Filmset and printed in Great Britain by
BAS Printers Limited, Wallop, Hampshire

British Library Cataloguing in Publication Data

Ransom, Philip John Greer
Your book of canals.
1. Canals – Great Britain – History
I. Title
386'.46'0941 HE435

ISBN 0-571-10971-3

Contents

Illustrations

Cover illustration: a narrow-boat style hire cruiser on the Birmingham & Fazeley Canal near Hopwas

ACKNOWLEDGMENTS OF ILLUSTRATIONS

Aerofilms Ltd: 6, 39, 56; British Waterways Board: 4, 36; D. Calkin: 3, 34, 37, 45; Deegan Photo Ltd: 22; Lound Hall Mining Museum: 10; Mary Evans Picture Library: 16, 17, 18, 29, 40, 41; Manchester Ship Canal Co.: 7, 8, 9, 46; National Library of Ireland: 13; North West Museum of Inland Navigation Ltd: 54; E. A. Ransom: cover illustration, 12, 24, 25, 47; Times Newspapers Ltd: 21; D. Salt: 15, 19; Union Canal Carriers Ltd: 32, 33, 50; Waterways Museum (BWB): 20, 26; Wiltshire Newspapers Ltd: 57; Author's collection: 11; Author: 1, 2, 5, 14, 23, 27, 28, 30, 31, 35, 38, 43, 44, 48, 49, 51, 52, 53, 55, 58.

1 *Canals Today*

A canal is a man-made waterway, a sort of highway on which the surface is water and the traffic is boats. Most canals in the British Isles were built between 1730 and 1830, and their purpose was to enable boats and barges drawn by horses to carry goods from place to place, or, in some instances, to enable ships to go inland. In those days the other main forms of inland transport for goods were either horse-drawn waggons, jolting over very rough roads, or packhorses. A single horse could pull a barge with a load weighing more than fifty tons; on a road, a horse could move only two tons. The advantage of canals was clear, and they provided the transport which made possible the Industrial Revolution.

Today, times have changed. Most goods are transported by motor lorry or railway train, and only in North-East England are goods much carried by barge. Canals now have many other uses. They are used to supply water to factories and farms. In the country, they help to provide land drainage, without which some farmland would be excessively boggy. In towns, they provide a source of water for fire-fighting, and carry away rainwater from roads. Everywhere they provide a great many anglers with waters in which to fish. Because many canals have been little modernised over the past century or more, they and their equipment are now of great historic interest.

Above all, though, canals are themselves attractive, and more and more people spend their holidays boating on them. Canalling combines the pleasures of boating with the joy of the open road: to go where you will and stop, within reason, where you please.

There is plenty of scope for this, for there are still a great many canals inviting exploration. To give some figures, there are in the British Isles about 1,630 miles of canals now in use by boats – that is, about 1,400 in England, 30 in Wales, 90 in Scotland and 110 in Ireland. Exploring them all would take a long time, for speeds are slow. Canals were designed for vessels moving at the pace of a walking horse, and motor craft are usually limited to four miles an hour. This slow speed seems strange at first, but it soon becomes part of the enjoyment – it enables the traveller to take in his surroundings in a way which he cannot do when rushing past by car.

As the map opposite shows, most English canals form part of a system of connected inland waterways, which also includes navigable rivers such as the Thames, Trent and Severn. It is extensive: by river and canal, boats can cruise from suburban Surrey northwards to rural Yorkshire, or from the mountains of the Welsh border eastwards to the flat lands of the Fens.

This system can be considered – as the map shows – in two

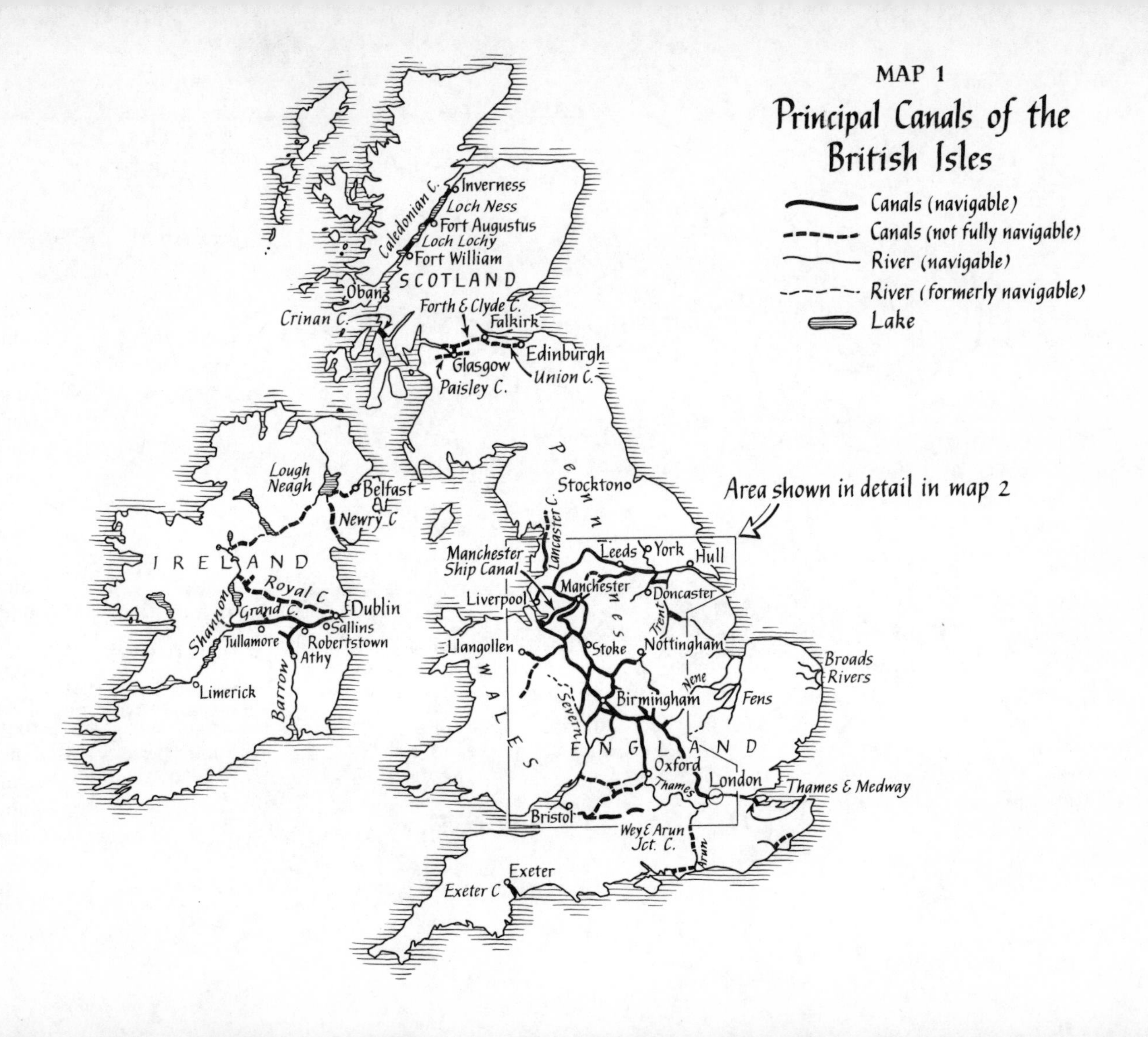
MAP 1
Principal Canals of the British Isles
Canals (navigable)
Canals (not fully navigable)
River (navigable)
River (formerly navigable)
Lake
Area shown in detail in map 2
Inverness
Caledonian C.
Loch Ness
Fort Augustus
Loch Lochy
Fort William
SCOTLAND
Oban
Crinan C.
Forth & Clyde C.
Falkirk
Edinburgh
Glasgow
Union C.
Paisley C.
Lough Neagh
Belfast
Newry C
IRELAND
Royal C
Dublin
Grand C.
Sallins
Shannon
Tullamore
Robertstown
Athy
Barrow
Limerick
Pennines
Stockton
Lancaster C.
Manchester Ship Canal
Leeds
York
Hull
Manchester
Doncaster
Liverpool
Trent
Llangollen
Stoke
Nottingham
WALES
Severn
Nene
Birmingham
Fens
Broads Rivers
ENGLAND
Oxford
London
Thames
Thames & Medway
Bristol
Wey & Arun Jct. C.
Arun
Exeter
Exeter C

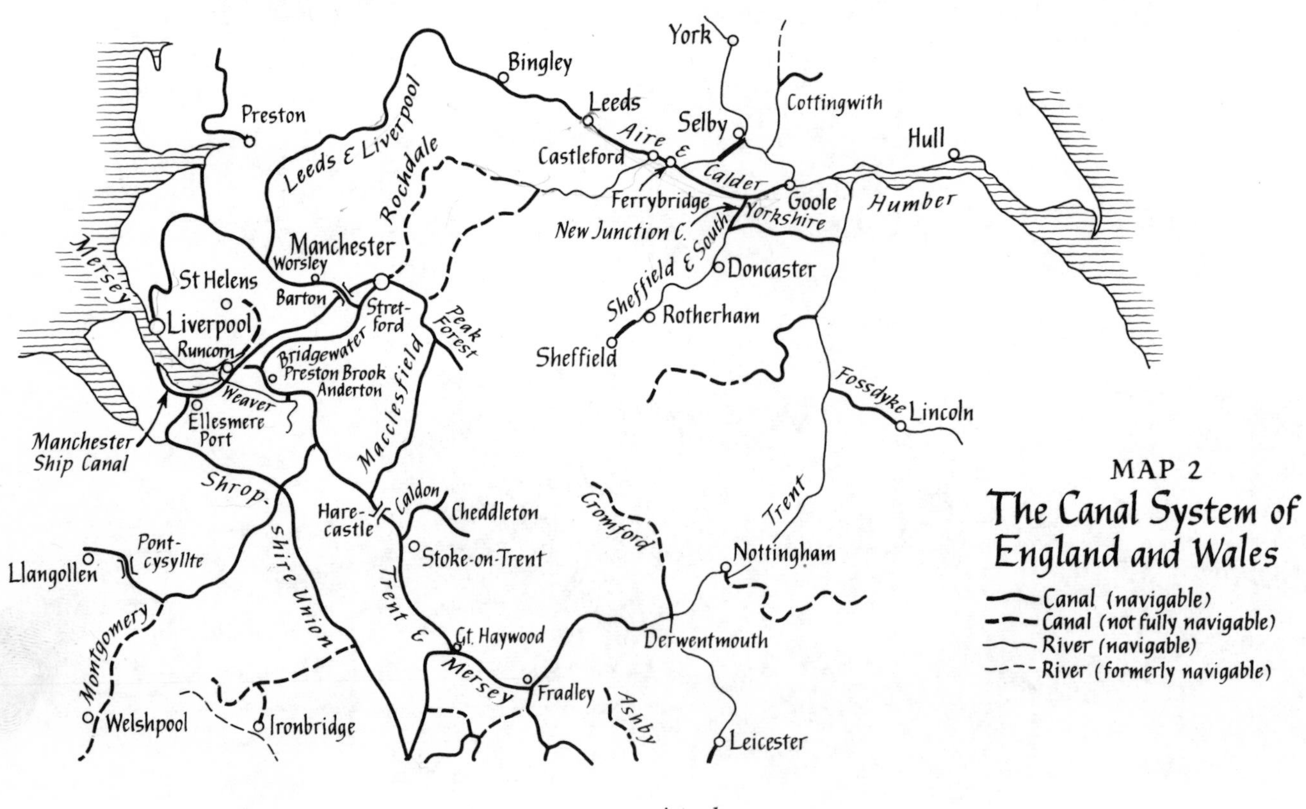
York
Bingley
Preston
Leeds
Cottingwith
Selby
Hull
Leeds & Liverpool
Rochdale
Castleford
Aire & Calder
Ferrybridge
Goole
Humber
New Junction C.
Yorkshire
Sheffield & South
Mersey
Manchester
Worsley
St Helens
Doncaster
Barton
Stretford
Peak Forest
Liverpool
Rotherham
Runcorn
Bridgewater
Preston Brook
Anderton
Sheffield
Weaver
Macclesfield
Fossdyke
Lincoln
Ellesmere Port
Manchester Ship Canal
Shropshire Union
Harecastle
Caldon
Cheddleton
Cromford
Trent
Pontcysyllte
Llangollen
Stoke-on-Trent
Nottingham
Montgomery
Trent & Mersey
Gt. Haywood
Derwentmouth
Fradley
Ashby
Welshpool
Ironbridge
Leicester
MAP 2
The Canal System of England and Wales
Canal (navigable)
Canal (not fully navigable)
River (navigable)
River (formerly navigable)

continued

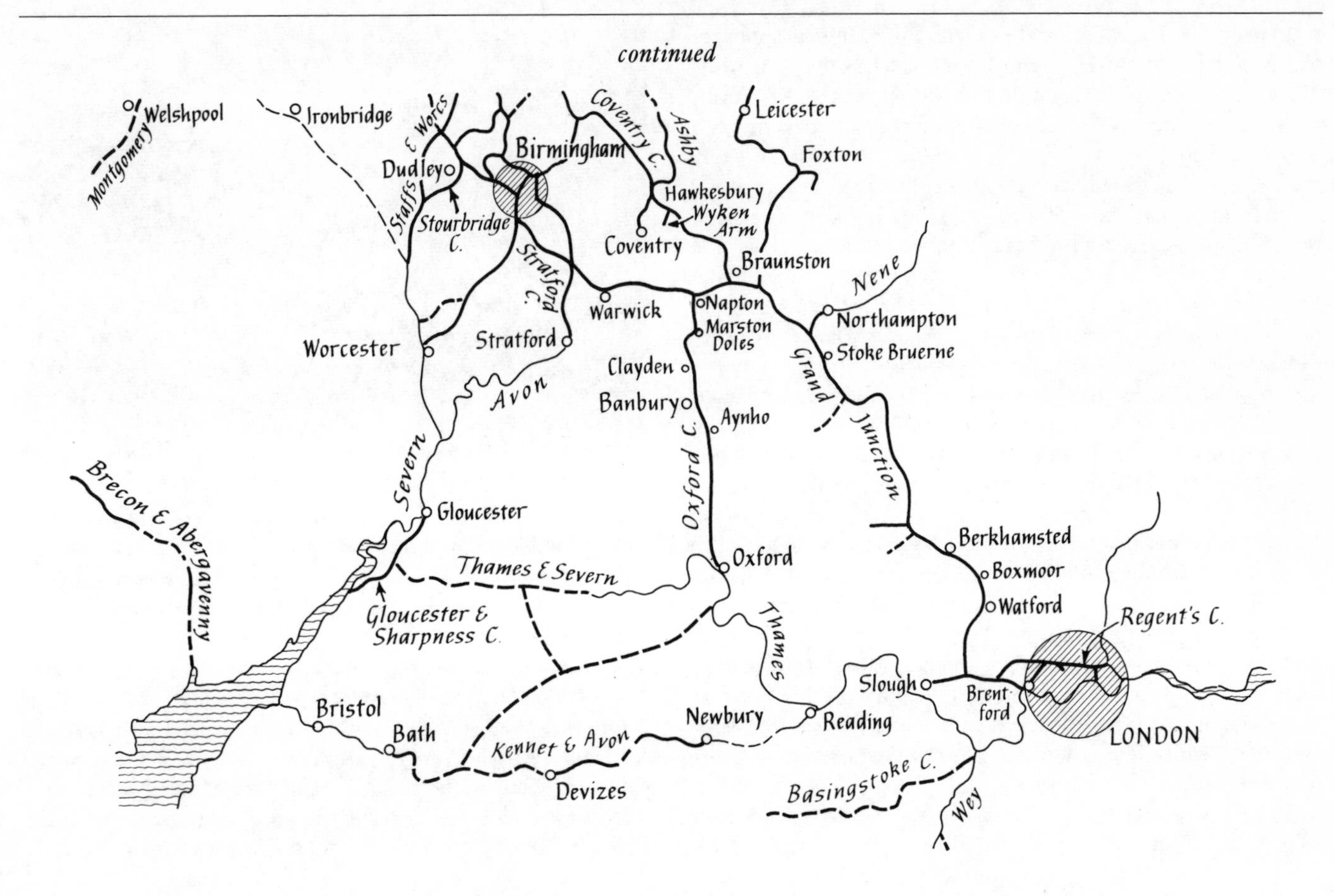
continued
Welshpool
Montgomery
Ironbridge
Staffs & Worcs
Birmingham
Dudley
Stourbridge C.
Coventry C.
Ashby
Leicester
Foxton
Hawkesbury
Wyken Arm
Coventry
Braunston
Stratford C.
Warwick
Napton
Marston Doles
Nene
Northampton
Stoke Bruerne
Grand
Junction
Worcester
Stratford
Clayden
Banbury
Aynho
Avon
Severn
Oxford C.
Brecon & Abergavenny
Gloucester
Thames & Severn
Oxford
Berkhamsted
Boxmoor
Watford
Gloucester & Sharpness C.
Thames
Regent's C.
Slough
Brent-ford
LONDON
Bristol
Bath
Newbury
Reading
Kennet & Avon
Devizes
Basingstoke C.
Wey

main parts. Firstly, there are the canals of the Midlands, most of which are narrow canals, intended for boats less than 7 ft wide. They link the industrial cities they were built to serve – London, Birmingham, Manchester and so on – but they also pass through much attractive countryside in between. It is remarkably attractive, considering how close much of it is to built-up areas.

On these canals is to be found the typical canal scene: between the meadows, a reed-fringed lane of water curves along the contour of a hillside, accompanied by grassy towpath and bushy boundary hedge. It is spanned by humped bridges of mellow red brick, and bears *narrow boats* with long black hulls and gaily decorated cabins. The traveller along such a canal encounters, sooner or later, locks, with massive wooden gates and sheer brick sides, by which the canal goes uphill or down, black tunnels by which it pierces high ground, and lofty aqueducts by which it crosses valleys; and he finds cuttings, embankments, draw-bridges, elegant eighteenth-century buildings, and junctions with other mysterious canals each as enticing as the first.

Eventually, though, the canal reaches a town. In a few towns and cities canals were laid out so that they added to the amenities – in London, the canal at Little Venice, and through Regent's Park, is an example. More often, though, canals were hidden away, at the backs of factories and behind high walls, like much of Birmingham's extensive canal system. Such canals are often dirty and strewn with rubbish, but they remain essential cruising links between rural canals.

They have other compensations, too. The typical by-pass factory which presents a showy facade to motorists offers

1. Humped bridge and narrow lock. Cropredy, Oxford Canal

canallers at its back door a less pretentious view. As you cruise by, open doorways reveal strange industrial processes – the lightning-flashes of a welding department give way to pungent aromas from a paintshop, and people in overalls look up from their work to wave.

These pictures are repeated elsewhere with variations. In North-East England the waterways are wider and the locks traditionally wide but short – although many have been enlarged for big barges which continue to carry oil or coal in that part of England. Here is the second main part of the system – the waterways which radiate from Hull and, part-river, part-canal, link the Humber ports with industrial towns such as Leeds,

2. Amenity canal: pleasure cruisers, a pair of camping boats and a narrow boat used as a floating canal-ware shop on the Grand Union Canal in 1975

3. Ornamental stonework on Huband Bridge over the Grand Canal in Dublin

Doncaster, York and Nottingham. Near Nottingham the River Trent joins the Midlands canals; and the Leeds & Liverpool Canal crosses the Pennines to make another connection with these canals near Manchester.

In Ireland, the scene changes slightly again. For instance, on the Grand Canal, the humped bridges are built not of brick but of grey limestone – in keeping with the softer colouring of the landscape generally. Grand Canal bridges usually have their names and eighteenth-century building dates carved in stone in old-fashioned letters, in place of the iron number-plates common on the English canals. This canal too is wide compared with most English canals; it connects the capital, Dublin, with the principal navigable rivers, the Shannon and the Barrow.

All these canals are only wide enough and deep enough to carry barges or boats. Where they reached the coast, goods were transferred to ships large enough to go to sea. Ship canals

4. Canoeists use the Padding Arm of the Grand Union Canal

5. Coal by canal: Doncaster power station, June 1975

6. The Caledonian Canal follows the contours along the side of the Great Glen and enters Loch Lochy in the distance

7. A large vessel from South Africa arrives in Manchester by ship canal

overcame this problem; the most important is the Manchester Ship Canal, opened in 1894, long after most canals, and built so wide and deep that ocean-going ships can reach Manchester, which is 60 miles from the open sea.

The principal canals in Scotland were also built as ship canals. The Caledonian Canal connected East and West Coasts so that sailing ships could avoid the dangerous passage round the North of Scotland; and the Crinan Canal, built early, and on a comparatively small scale, is a short cut for boats going between the Clyde and the West Coast. The Caledonian Canal in particular is a fine place to cruise; some of the highest mountains in Britain stand close to it, and its course includes several large lochs and takes in Loch Ness, so there is a chance to look out for monsters.

As I mentioned earlier, all these canals (except the Manchester Ship Canal) have lost much or all of their original purpose. How this came about is described in the next three chapters.

2 *How Canals Grew*

Rivers have been used as routes for transport since prehistoric man first floated down one on a tree trunk, and by the Middle Ages they were used extensively. But rivers tend to flood after winter rain and snow, to become very shallow from summer droughts, and to consist at all times of shallow, fast-flowing sections alternating with deeper, slower-flowing ones. This made them difficult to navigate. During the seventeenth and eighteenth centuries many rivers were improved for navigation, or made navigable where they had been un-navigable before. Weirs were built to hold back water so that it would be deep enough for barges, and locks were built for barges to pass the weirs. To link the locks with the original river, lengths of artificial waterway were cut, ranging in length from a few yards to several miles. In some places these are called canals, particularly in Ireland.

Canals in Britain were first built by the Romans, and the Fossdyke, in Lincolnshire, is a survivor. In Devon, a short canal, 1¾ miles long, was opened in 1566 to enable vessels to reach Exeter from the estuary of the River Exe. Canals had been built on the Continent since the Middle Ages. However, the great era of canal construction in the British Isles, which lasted for about a century, commenced in Ireland in 1731 when work started on building the Newry Canal. This was opened in 1742; it was 18½ miles long and linked Lough Neagh with the Irish Sea. It was the first result of extensive schemes for canals in Ireland and it was followed, eventually, by many other canals. Work started on the Grand Canal about 1756. This canal was opened in sections: not until 1804 was its main route complete through to the Shannon.

By then the canal age had long since spread to Great Britain. A proposal to make the Sankey Brook navigable from St Helens, Lancashire, to the River Mersey had resulted in what was virtually an entirely artificial canal, opened in 1757. The big leap forward came, however, with construction of the Bridgewater Canal. In the 1750s the Duke of Bridgewater (who had seen canals in Europe) owned coal-mines at Worsley, not far from St Helens. He wished to sell coal in Manchester, a little over 10 miles away to the south-east, and to do so he needed cheap transport. To achieve that, he decided to have a canal built.

The engineer who designed and laid out the canal was James Brindley.

Brindley, though scarcely able to read and write, was a practical genius and had acquired a great reputation as an engineer. He was already 43 years old when he started work on the Duke of Bridgewater's Canal in 1759. His early training had been as a millwright; he had gained much experience in

8. Francis, third Duke of Bridgewater

9. James Brindley

10. At Worsley the Duke of Bridgewater had an underground canal system built to serve his coal mine. These are the entrances, with boats moored outside.

construction of water courses and control of water.

The canal, which was opened in 1761, was, and still is, level all the way to Manchester, without locks. It achieved this only by crossing the River Irwell (itself navigable) by an aqueduct at Barton. No canal aqueduct had previously been built in Britain, and Brindley's proposal to carry one waterway over another seemed to many to be a 'castle in the air' – until it was successfully completed.

From this little canal, the canal network started to spread. The duke had decided to have canal access for his coal not only to Manchester but also to Liverpool, and at the same time to have a canal which could carry goods between those two cities. So from Stretford, 6½ miles from Worsley, a canal was built for some 26 miles to Runcorn, where it descended through locks to join the Mersey Estuary. A few miles short of Runcorn, at Preston Brook, it connected with another new canal, the Trent & Mersey or Grand Trunk Canal.

This was a far greater undertaking than the Bridgewater Canal, and Brindley had been surveying it even before he started on the Bridgewater. It was intended to link the Mersey in the west with the navigable Trent in the east. In doing so it would make a waterway link across England for the first time, and it would also improve transport to and from the Potteries, through which it passed. This would mean that coal and china clay could be brought in by boat, and completed china despatched by the same method – with fewer breakages than would occur with packhorse or road waggon. To jump ahead for a moment, an example of this particular advantage of canal over road survives today: Johnson Bros. have two factories a few miles apart beside the Caldon Canal, a branch of the Trent & Mersey, and they have found it worthwhile to build two boats during the past few years to move pottery between these factories.

The main line of the Trent & Mersey Canal runs from Preston Brook to the Trent above Nottingham, 93½ miles with 76 locks. Through hills which form the watershed near Stoke-on-Trent is a tunnel over 1½ miles long at Harecastle – or, to be precise, two tunnels, since a second tunnel was made in the 1820s and the original is now disused. Before Brindley started to bore it, few if any tunnels of any length had been made in Britain except in mines. It took 11 years to complete, and the Trent & Mersey Canal was not open throughout until 1777.

In two respects the T & M was unlike the Bridgewater, but like many canals built later in England and Wales. It was built, for most of its length, as a narrow canal; and it was built and run by a company incorporated for the purpose by Act of Parliament. The money needed to pay for building canals was provided, usually, by people who subscribed for shares in such companies. They did so partly in the hope – sometimes justified, sometimes not – of making a profit, and partly because they were public-spirited; they thought that canals would benefit the public. In Scotland and Ireland some canals were built by companies but others were built by the government, and throughout the British Isles the government sometimes loaned money to companies building canals.

Just as the Trent & Mersey was a larger undertaking than the Bridgewater, so it itself formed part of a still greater plan put

11. Brindley's Barton Aqueduct carried the Bridgewater Canal across the River Irwell

forward by Brindley. This was for *the cross*: a canal system in the shape of an X which would link not only Trent and Mersey but also Severn and Thames as well. The T & M, which runs south-east from Preston Brook almost to Fradley, Staffordshire, before turning north-east towards the Trent, forms the two upper arms of the cross. From Great Haywood, a few miles short of Fradley, the Staffordshire & Worcestershire Canal was built as the south-western part of the cross to meet the Severn at Stourport; and from Fradley Junction canals were gradually built over the route to Coventry and on via Banbury to the Thames at Oxford.

These early canals were very successful. They carried much traffic, benefited the places they served, and earned money for their shareholders. Even while they were still being built, other canals were being promoted elsewhere. Places near the canals of the cross had canals built to link them to it. In the north, three east-west canals were built across the Pennines, and in the south, east-west canals linked Thames and Severn, and Kennet (a tributary of the Thames) and Bristol Avon. The Wey & Arun Junction Canal, joining the rivers of its name, linked London with the South Coast. Other canals filled in the gaps – notably the Grand Junction Canal, which was built from London to Braunston, Northants, where it joined the Oxford Canal. The result was a more direct route between London and the Midlands than that via the Thames.

Most canals formed part of the main system but there were others which were isolated from it, such as the canals of the South Wales valleys, which linked collieries with ports.

Canals in Scotland, too, were separate from the English ones, and to a large extent from each other, though a small network did grow up in the central Lowlands. The first part of this, and the first canal in Scotland of any importance, was the Forth & Clyde Canal, which was opened in sections between 1775 and 1791. Its main line was 35 miles long and it linked the firths and estuaries of its name. There was a branch to Glasgow and the whole canal was built large enough to take small ships. Later the Union Canal (or Edinburgh & Glasgow Union, to give it its full name) was built from Edinburgh to join the Forth & Clyde near Falkirk.

The Caledonian Canal runs from Clachnaharry, near Inverness, to Corpach, near Fort William. It was intended to make it large enough for full-size ships of its time, but when, after many years in building, it was opened throughout in 1822, it was still incomplete, and had to be deepened and widened in the

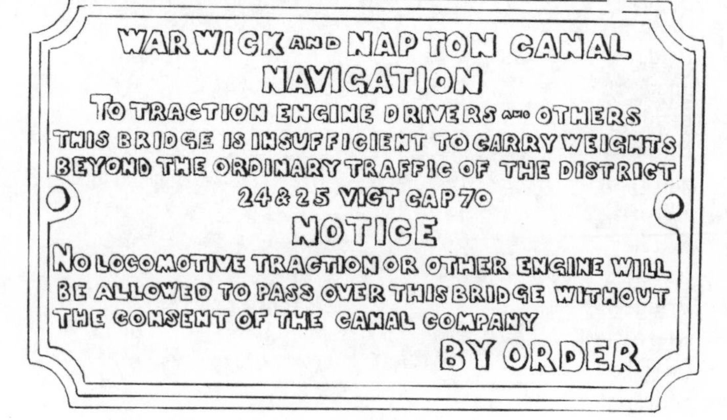

12. Canal bridges were built for horses and carts, so in Victorian times notices like this became common

1840s. This canal was built and administered by government commissioners, and its engineer was Thomas Telford.

Telford designed and built canals in England, too, and although many great canal engineers followed Brindley (Watt, Jessop and Rennie are famous), of them all Telford was the greatest. That the Caledonian was incomplete when opened was not his fault.

I have pointed out that the Scottish canals were mostly separate from each other, but the Crinan Canal (opened in 1801), the Caledonian and to a lesser extent the Forth & Clyde did become components of the network of water communications along the West Coast of Scotland and to and from its islands.

What did all these canals carry?

Canals were intended mainly for transport of goods in bulk. Of these, transport of coal was the reason for building not only the Bridgewater Canal but many others also. Other important bulk goods carried by canals were limestone (for making roads, and for turning into fertiliser), bricks, slates, timber, hay, corn and cotton. In addition, Scottish canals carried cattle and herrings, and Irish canals peat and porter. Canals were busy, too, carrying all sorts of parcels and general merchandise.

They also carried passengers. People had for a long time travelled by river; in 1772 passenger boats started to run on the Bridgewater Canal. Travel by boat was a smooth, comfortable alternative to travel by road, whether on horseback, on foot, by carriage or by waggon. Canal passenger boat services became widespread: to give a few examples, on the Grand Junction

13. A passenger boat passes barges on the Royal Canal

Canal between Paddington and Uxbridge, on the Kennet & Avon Canal near Bath, between Preston and Kendal on the Lancaster Canal and between Glasgow and Edinburgh. In Ireland there were extensive passenger services over the Grand Canal, with grandiose canal-side hotels built to accommodate passengers at the end of each day's journey.

All these services used horse-drawn boats. However, the first successful steamboat service in Europe started on the Firth of

Clyde in 1812 and by 1819 the original vessel, the *Comet*, was running between Glasgow and Fort William, passing through the Crinan Canal on the way. Steamers were operating on the northern part of the Caledonian Canal before the rest of it was finished, and when it was all open they started to run between Glasgow and Inverness, passing through the two canals.

Of all the vessels carrying goods or passengers on canals, only some of the horse-drawn passenger boats were operated by the canal authorities themselves. All other carriage on canals was done at first by independent carriers – the best-known name from the early days is that of Pickfords. Canals were considered to be comparable to the turnpike roads of the period. The canal company, or other authority, provided the canal, and it was open for anybody to take a boat along it, provided the size of the boat was suitable and the tolls were paid. This 'public right of navigation' was granted by Parliament in Acts which set up canal companies, in return for rights given to the canal companies themselves – such as to be able to buy, compulsorily, the land they needed to build their canals.

14. The Union Canal near Linlithgow: canoeists use stables (left) built for passenger-boat horses

3 *Canals in Decline*

Stockton-on-Tees, though now part of the large built-up area called Tees-side, retains the broad main street of a country town, with the town hall standing in its centre. Not long ago the author attended a meeting in it, and learned that in the very same room had been held a meeting in the far-off 1820s at which a historic decision was made: that Stockton should be linked with coal-mines to the west not (as had been planned) by a canal, but by a railway.

The railway became the Stockton & Darlington Railway, and with its opening in 1825 the railway age began. Canals became less important than before.

There had been railways, of a sort, for many years. These railways, or tramroads, were simple, primitive affairs on which small waggons were pulled by horses. It cost more to carry goods by tramroads than canals, but they were, in hilly districts, cheaper to construct. Often they were built to link mines or quarries with nearby canals and in some places they formed part of a route which also included canals. The northern part of the Lancaster Canal remains open but isolated because the tramroad which once connected it with the main canal system has long since been closed.

Some early, experimental steam locomotives were tried out on canal tramroads. On the Stockton & Darlington Railway ran improved versions, and still better ones on the Liverpool & Manchester Railway, opened in 1830 in direct competition with the Bridgewater Canal. Steam railways appeared to offer a cheaper, faster means of transport than canals, and within a few years they were being built in most parts of the British Isles. Like canals, railways were built and run by companies, but the railway system expanded even faster than the canals had.

All the same, two important canals were opened in the 1830s: the Macclesfield in 1831, and the Birmingham & Liverpool Junction in 1835. Canals did lose their horse-drawn passenger traffic soon after competing railways were built – the Edinburgh & Glasgow Railway was opened in 1842, for instance, and the passenger boats between those two places were withdrawn in 1848. Even after railways were open some canals continued to carry bulk goods. They could only do so, however, if they reduced their tolls, and the carrying firms reduced their rates. This made canals less profitable, and less able to afford improvements.

In 1845 an Act of Parliament authorised canal companies to set up their own carrying fleets, and many canal companies did so. They included the Grand Junction, the Leeds & Liverpool, the

Shropshire Union (formed by amalgamation of the Birmingham & Liverpool Junction and other canals) and the Grand Canal.

Much was heard, at the time, of converting canals into railways: but in a few instances only was this done. One of them was Strood Tunnel. Today it is used by trains on their way from London to North Kent, but it was built for the Thames & Medway Canal. In many instances, however, canals were bought by railway companies: new railways wanted to control their main competitors, while canal shareholders sold while they could. Some railways maintained and even developed the canals they owned, particularly if this enabled them to draw traffic from places not served by their own railway routes. The North Staffordshire Railway, which purchased the Trent & Mersey Canal, is a good example.

15. Consall on the Caldon Canal, 1924: railway and canal were both owned by the same company

More often, though, railways neglected their canals and discouraged people from using them by raising the tolls. The most notorious example was the Kennet & Avon Canal, owned by the Great Western Railway. Eventually so many canals in the main English network had been purchased by railways that they split apart those canals which remained independent. As a result, Parliament passed Acts during the middle of the nineteenth century to oblige railway companies to keep their canals navigable for anyone who wished to use them.

Towards the end of the century, people began to favour canals again. Some waterways, notably the Aire & Calder, were improved, and some new sections of canal were opened – the canal to Slough, for instance, in 1883 and the New Junction Canal, which links the Sheffield & South Yorkshire and the Aire & Calder Navigations, in 1905.

Far more important was the Manchester Ship Canal. In the middle of the nineteenth century Manchester had been a city declining in importance and prosperity. One of the chief causes was the very high charges imposed on goods imported or exported through the docks at Liverpool. A canal from the Mersey estuary to let ships come right up to Manchester was considered at intervals; the scheme that was eventually successful originated in 1882. It took until 1894 to complete the canal and during that time the new Manchester Ship Canal Company acquired the Bridgewater Canal and itself came under

16. Building the Manchester Ship Canal

the control of Manchester Corporation, which loaned sufficient money for the ship canal to be completed.

Though technically a canal, the Manchester Ship Canal is in effect an elongated port with quays and docks not only in Manchester but at intervals along its entire 36-mile length. It has been and is a successful undertaking: traffic has increased from 925,000 tons of goods in 1894 to about 16,000,000 tons a year at the present time, and about 5,000 ships enter the canal each year from many parts of the world.

Revival of interest in waterways at the turn of the century culminated with the appointment of the Royal Commission on Canals and Inland Navigations in 1906. The commission made a most comprehensive study of the inland waterways of the British Isles and in 1909 issued, in twelve volumes, its report and

17. Narrow boats enter a lock on the Grand Junction Canal

recommendations, which included enlarging many canals to take bigger craft.

Not one of the commission's recommendations, so far as I am aware, was ever adopted. Why not? Partly because those in favour of railways were too strong; partly because the first world war intervened; and partly because the internal combustion engine had been invented and was coming into use. It greatly altered the balance of advantage between each means of transport – water, rail and road.

At first, canals benefited from internal combustion engines. They were compact enough to be fitted into barges and boats. Steam barges and narrow boats had been built, but the steam engine and boiler took up so much space that they left little room for cargo. So most barges had continued to be pulled by horses. Internal combustion engines were successfully fitted to narrow boats and barges from the 1900s onwards and eventually superseded both steam and horse-drawn boats; though horses lasted a long time and even now, as I write, one firm of carriers continues to use a horse to move boats laden with rubbish along the canals of Birmingham.

However, the internal combustion engine gave road traffic a great advantage over canals, and over railways too. Motor lorries, people discovered, could carry goods quickly from door to door rather than from wharf to wharf or siding to siding, and motor buses replaced surviving passenger services on canals such as the Crinan and the Gloucester & Sharpness which had been worked by steamers.

There were occasional attempts to modernise canals. For

18. A Victorian view of canals

instance, on the main route from London to Birmingham (over which the Grand Junction and several other companies had amalgamated to form the Grand Union Canal) narrow locks from Braunston to Birmingham were replaced by wide ones during the 1930s. This produced a wide canal from London to Birmingham, since the Grand Junction already had wide locks; traffic, however, continued to be worked by narrow boats in pairs, passing through the locks together. On the whole, such canal modernisation as was done was too little and too late, and the period between the wars was one of continuing decline for canals.

19. Horse-drawn boats were busy on the Caldon Canal in 1910

20. Unloading narrow boats at Birmingham during the 1950s

4 *The Canal Revival*

Canal companies in Great Britain were nationalised in 1948, along with railways and many road haulage firms. During the second world war the government had controlled those canals still considered important for transport, and it was these that were nationalised (the Manchester Ship Canal, considered as a port, being a notable exception). Also nationalised were those canals owned by railway companies. Some of these were still used, but others were almost derelict (despite Acts of Parliament of the previous century) and still others had been closed to navigation but were kept in existence for purposes such as water supply. A few canals which were no longer considered important for transport, such as the Rochdale and Basingstoke Canals, were not nationalised.

The authority which owned and ran nationalised transport was, at first, the British Transport Commission. Later, in 1962, the commission was disbanded and the British Waterways Board, with its members appointed by the government, was set up to run the nationalised waterways. Although nationalised, individual canals are still called by their old names. Canal carriers' boats and barges were not nationalised, except for the canal companies' own carrying fleets, though one of the biggest independent fleets, that of Fellows Morton & Clayton, was sold to the BTC soon after nationalisation.

Nationalisation did not affect canals in Northern Ireland, but their life was nearly over anyway. The Newry Canal, disused after 1939, was formally closed in 1949, and most other canals in Northern Ireland were closed by 1951.

In the Irish Republic, the Grand Canal Company was merged in 1950 with Coras Iompair Eireann, the national transport organization which runs railways and road services. The canal company had been a pioneer of diesel barges, but it had also built up a fleet of road lorries. CIE ceased to carry by barge in 1960 and, since independent canal carriers had already ceased to operate, the Grand Canal has not been used for trade since then.

Had the Grand Canal Company been able to remain independent of CIE, it would probably have turned itself, eventually, into a road haulage business. In 1950, so its last general manager pointed out to me several years later, a barge with a crew of three, working 16 hours a day, could carry a load of 50 tons from Dublin to Limerick (the most important route which included part of the River Shannon) in 4 days. A lorry with a crew of one could carry 10 tons between the same places in a normal working day with a little overtime. Since then, lorries have been made which carry far bigger loads, while trades union pressure to reduce the hours worked by barge men would probably have increased the time taken for the journey by water.

Withdrawal of barge traffic in Ireland had a parallel in England when British Waterways Board ceased to carry by narrow boat about 1964. The very hard winter of 1962–3 meant that canals were frozen and boats immobilised, despite the efforts of icebreakers, so that traffic was lost to canals just at a period when both roads and railways were being modernised. There was one traffic which BWB did continue to carry by narrow boat: barrels of lime juice from Brentford (where it arrived by lighter from London Docks) to Boxmoor on the Grand Union Canal. This traffic still continues at the time of writing, though it is now handled by independent carriers, so anyone who wants to help keep goods moving by canal should drink lots of Rose's Lime Juice!

When BWB stopped carrying by narrow boat it was not the total end of working narrow boats, even apart from the lime-juice run, since several independent carriers remained. But many of them gradually closed down during the 1960s. One interesting new development has been the retail coal trade: people concerned about canals sell coal from their narrow boats, travelling round the canals to do so. Despite this, however, working narrow boats have become very rare.

This decline in trade has been matched over most inland waterways in Great Britain, but it is worth comparing them with those of the rest of Europe. Continental inland waterways have constantly been developed, and many of them are wide enough and deep enough for barges which carry loads as large as 1,350

21. Icebreaker at work on the Grand Union Canal in February 1963. The crew rock the boat to break the ice

tons. Recently the president of France announced that a *new* 127-mile canal of this size is to be built to connect the Rivers Rhône and Rhine. Barges which are considered small on the Continent carry loads of 350 tons. In England narrow boats carry 25 to 30 tons. It costs too much to move such small loads slowly by water.

However, where waterways are big enough to carry large loads they do so, in England as on the Continent. Barges carry 500 tons of petroleum at a time from the Humber ports up the Aire & Calder Navigation, and locks are being lengthened to admit 750-ton barges. The Gloucester & Sharpness Canal takes coastal tankers carrying 1,000 tons of oil from the Severn Estuary almost to Gloucester.

Unfortunately, unlike Continental countries, Britain is missing opportunities to enlarge canals, and lacks the benefit of cheap bulk transport that they bring. A case in point is the Sheffield & South Yorkshire Navigation. Barges carrying 500-ton loads can reach Doncaster from the Humber, but the largest barges that can go beyond that place carry only 90 tons. The British Waterways Board wishes to enlarge the waterway and locks so that 700-ton barges will be able to reach Rotherham, 15 miles above Doncaster. The government has yet, at the time of writing, to make the money available for it to do so.

Although freight traffic on the small waterways of the British Isles has declined sharply since the end of the second world war, there has been during the same period a surge of public interest in canals.

This originated from the publication in 1944 of the book *Narrow Boat* by L. T. C. Rolt. In it he described an extensive

22. A hack boat, or barge owned by an independent carrier, on the Grand Canal about 1950

cruise made over the canals of the English Midlands in 1939. Cruising on inland waterways was then mostly limited to the River Thames and the Norfolk Broads, and to cruise on canals, though not unknown, was most unusual. This was partly because many canal companies discouraged pleasure craft, fearing that they would get in the way of trading vessels, and partly because most people did not realise that the canal system existed. Certainly, people knew of canals: if you lived in a town, the canal was a smelly strip of water past the gasworks; if you lived in the country, it was a pleasant strip of water to walk beside with the dog; if you motored you knew that from time to time the car would lurch over a hump-backed canal-bridge; but it was from *Narrow Boat* that many people (including the present author at the age of fourteen) discovered that there was a network of canals, entire yet secretive, by which you could explore a very large part of England, one which had, moreover, altered little for 100 years. The hobby of cruising on canals has been increasing in popularity ever since.

Another effect of *Narrow Boat* was the formation of the Inland Waterways Association. It was founded by Robert Aickman, who was already interested in waterways when he read the book. He then joined forces with Rolt to form the IWA in 1946. The association has been campaigning ever since for retention and development of inland waterways in the British Isles and for their full use for commerce and pleasure. Its formation has been followed by the formation of many local waterway societies with interests in particular canals and rivers, and by the Inland Waterways Association of Ireland in 1954 and the Scottish Inland Waterways Association in 1971. The IWA and many related groups offer junior membership.

Despite the campaigns of voluntary organisations, much commercial traffic has left canals, for reasons already explained. As I write, the IWA is campaigning strongly in the cause of inland shipping: large barges on wide, deep canals able to carry big loads economically.

The Inland Waterways Association and similar bodies have been very successful, however, in making people and politicians aware that existing small canals have many uses – those that I mentioned at the beginning of chapter one – and that they are valuable amenities. In support of their campaigns, many waterway societies hold rallies of boats on little-used canals. From far afield, pleasure and trading craft converge on the rally site, and people living nearby are able to see the canal's potential for themselves.

Following the change in public opinion, Parliament, in the Transport Act 1968, classified the waterways (both canals and river navigations) of the British Waterways Board thus:

Commercial Waterways – about 300 miles, to be principally available for carriage of freight;

Cruising Waterways – about 1,100 miles, to be principally available for cruising, fishing and other recreational purposes;

The remainder – about 600 miles, of which less than 250 miles were then navigable, to be dealt with in the most economic manner possible.

23. Many former commercial wharves have become pleasant bases for pleasure craft. This one is at Aynho on the Oxford Canal

Unfortunately the same Act abolished the public right of navigation over the Board's waterways. Were it not for the activities of the Inland Waterways Association, it is most unlikely that the 1,100-mile network of cruising waterways would have survived.

Strictly speaking, the waterways of the BWB system belong to the Board. However, its chairman Sir Frank Price has expressed this view: that the Board only manages the system – which really belongs to the people.*

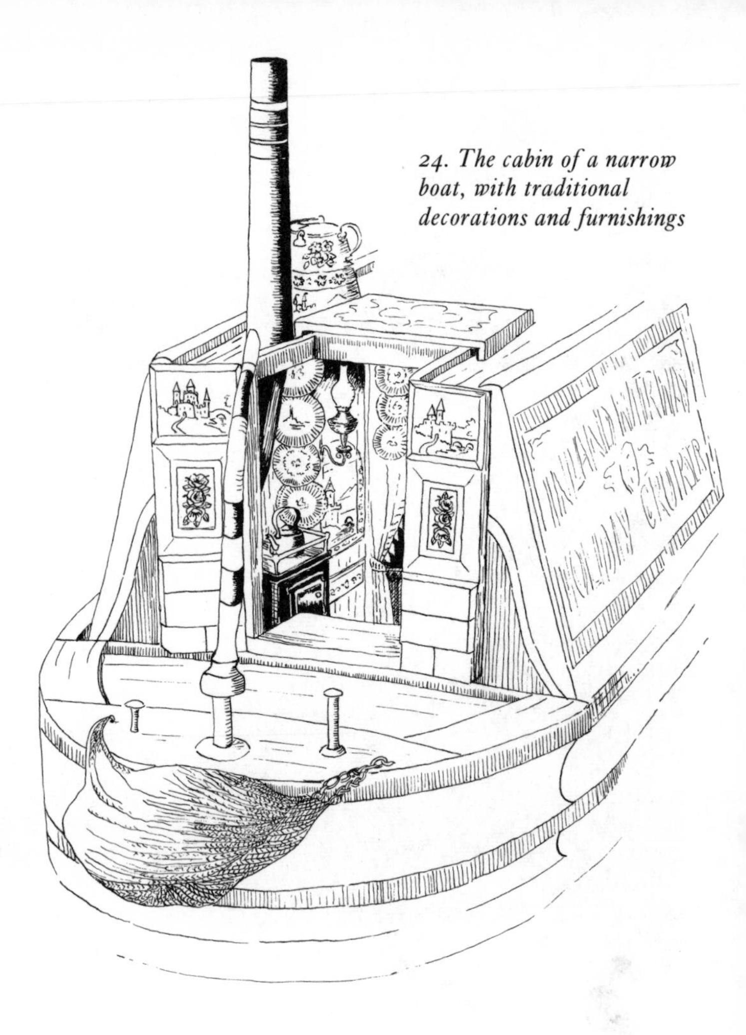

24. The cabin of a narrow boat, with traditional decorations and furnishings

* Since the above was written, the government has suggested that a National Water Authority should be set up and the BWB System merged into it, except for BWB waterways in Scotland which would pass to a Scottish administration.

5 *Boats and Barges*

25. A narrow boat's traditionally-decorated water cans and mop

The largest boat that could go comfortably almost anywhere on the main English inland waterway system would be 45 ft long and 6 ft 10 in. wide; it would draw about 2 ft of water and its greatest height above the water line would be about 5 ft 6 in.

No commercial craft were built within these dimensions for this purpose, though many pleasure craft now are. Some early canals were built to take barges already in use on adjoining rivers and estuaries; later barges and boats were built to fit the canals over which they were intended to work. Length and beam (that is, width) of craft usually depended on the length and width of the locks, draft on the depth of the canal and height on the bridges and tunnels.

Canals can be grouped according to their dimensions, from smallest to largest: tub-boat canals, narrow canals, wide canals and ship canals.

Tub-boat canals were a development in the early days. There are no tub-boat canals now in use, though a short section of one is preserved in the Blists Hill Open Air Museum, Ironbridge, Shropshire. The boats were very small – about 20 ft by 6 ft; they were towed by horses in trains of twenty or so. One of the original tub boats of the Shropshire Canal has been re-floated on the preserved section: it was discovered on a farm in use as a water tank, and that is just about what it looks like! It is most interesting, though, as a surviving example of a form of transport now extinct.

The sort of boat most typical of English canals is probably the narrow boat built to fit the 72 ft by 7 ft locks of the narrow canals of the Midlands. Early boats were built entirely from timber, later ones sometimes of timber but often of iron or steel or as composites with steel sides and timber bottoms.

In the early days narrow boats were probably entirely open or perhaps had small cabins like the *day boats* still to be seen on the canals around Birmingham. Probably they were worked by crews of men who lived ashore, and only when railway competition forced boatmen to economise did they take their families to live on board to help work the boats. The most striking feature of family narrow boats is their traditional 'roses and castles' decoration. Against a plain but bright background (often red or green) are gaily painted circles, diamonds, geometrical shapes and bunches of flowers. Panels carry landscapes of fairy-tale castles, hills, lakes and bridges, and on the cabin-side appears in ornate letters the name of the owner of the boat, which may be a company large or small or an owner-boatman. The origin of the traditional painting is now unknown. There is an attractive theory that boatmen's families, when they moved from land to water, had to leave their gardens behind – and so painted their boats with flowers.

The living cabin of a narrow boat is not large – only about 10 ft long – and the layout became as traditional as the decoration. The cabin is at the stern of the boat, with doors opening from the counter – the platform where the steerer stands. Inside, on the left, is a coal stove; beyond is a cupboard of which the front folds down to form a table. At the far end of the cabin is a bed across the width of the boat, and on the right-hand side a bench seat which doubles as a child's bed at night. The water supply consists of a couple of traditionally painted cans kept on the cabin roof next (always) to the stove chimney which (always) has bands of brass round it for decoration.

When motor boats were introduced, the motor was placed forward of the cabin, so, since motor boats usually towed a 'butty' (an un-powered boat of horse-drawn type) a family usually had two cabins. Even so they did not have a lot of space, but these little cabins were home to generations of people. Life on a canal boat, though cramped, was I suspect preferable to living in a slum house in the mid-nineteenth century, and probably healthier. By the standards of the mid-twentieth century it lacked normal amenities such as mains electricity and running water and was less acceptable. A further problem was the difficulty of ensuring that children of boat families were properly educated, or even learned to read and write. They had to go to school wherever they could, a day or two here, a day or two there; and in some places special schools were set up for them.

On wide canals have worked many kinds of wide boats and barges. A few, such as the wide boats which used to work on the southern part of the Grand Junction Canal, were family boats, but most were not.

The Leeds & Liverpool Canal has *short boats* to fit its 62 ft by 14 ft 3 in. locks but, as with narrow boats, there are very few now trading regularly. Steam short boats ran between the 1880s and the 1950s and were the most successful small steam commercial craft on canals. 'Sheffield size' barges are used on the Sheffield & South Yorkshire Navigation, where the locks are 61 ft 6 in. long by 15 ft 6 in. wide. This means that the barges are shorter, conspicuously, than narrow boats: they are short, wide and

26. Launching one of the last wooden short boats built for the Leeds & Liverpool Canal

27. Modern transport: a tug propels three compartment barges carrying a total of 480 tons of coal through a lock on the Aire & Calder Navigation

tubby, having the proportions of toy boats that small children play with in the bath. Their saving grace is not so much width as draft. They have surprisingly deep holds which enable them to carry the 90-ton loads which are still – just – economical at the time of writing. The waterway is deep enough for boats of 6 ft draft. Most small canals were designed for boats drawing about 3 ft 6 in., but silting now often reduces even this.

Irish canal boats or barges were short and wide, too, by English standards. Those of the Grand Canal were about 61 ft by 13 ft – or rather are, for some survive, and are used to carry materials for maintenance of the canal.

Large modern barges in Great Britain are to be seen mainly on the Aire & Calder and the Sheffield & South Yorkshire up as far as Doncaster. They are often as long as 130 ft or more, and 18 ft 6 in. wide – that is, almost twice as long as a narrow boat, and more than two and a half times as wide. Some carry coal, sand, or general merchandise, and others are tankers which carry oil.

More striking, however, on these waterways are the compartment boats. As long ago as 1865 W. H. Bartholomew developed the early tub-boat idea into the compartment boats which became known as 'Tom Puddings'. The system remains in use. Each boat is, in effect, a floating rectangular box which holds 35 tons of coal, or two juggernaut lorry loads. They work in trains of twenty or so, towed by a single tug and close-coupled so that they snake round bends. They are loaded near collieries and towed to the port of Goole, where each compartment boat is lifted out of the water by a hoist for its contents to be tipped into a waiting ship.

A modern version of the compartment boat system was introduced in 1967 to supply coal to an electricity generating station at Ferrybridge. Carriers Cawoods Hargreaves Ltd use specially built barge trains in which a tug either pulls or (usually) pushes three compartment barges – but each compartment is much larger than the old type; it carries about 160 tons of coal, so the capacity of the train is about 480 tons. Nevertheless each compartment is hoisted when it arrives at the generating station to tip its contents out.

Another recent development which uses compartment barges is the Bacat system. *Ba*rge aboard *cat*amaran, it means: barges can go direct from places on inland waterways in England to places on inland waterways on the Continent because the barges themselves are carried across the North Sea on a mother ship of catamaran (twin hulled) type. Bacat barges carry 140 tons each and are pushed or pulled in trains of three or more by special tugs. The Bacat ship service between Hull and Rotterdam started in 1974, but at the time of writing (early 1976) it is not operating. Dockers at Hull, concerned at losing the work of loading and unloading ships, have persuaded their counterparts elsewhere not to load or unload Bacat barges.

The types of vessel to be seen on ship canals depend again on the size of the waterway, from the ocean-going ships on the Manchester Ship Canal to the coasters which pass through the Caledonian and Crinan Canals. The latter two are also much used by fishing boats on their way between home port and fishing grounds. The Crinan Canal was formerly much used by Clyde Puffers, small steamers which carried cargo between Glasgow

28. The Caledonian Canal is used by fishing boats travelling between West and East Coasts. Here one leaves the bottom lock at Fort Augustus, to enter Loch Ness in the distance

and the West Coast and Islands of Scotland; they were made to fit the locks of the Crinan and Forth & Clyde Canals. A few still trade, though powered by diesel engines rather than steam.

The Crinan Canal, as I mentioned earlier, was also used by some of the very first passenger steamers. But as steamers were developed they became larger, too large to pass through the canal. Passengers were carried along it for some years by horse-drawn 'track boats', and had to change to or from steamers at each end. Then in 1866 the little steamer *Linnet* was built specially for the canal, to supersede the track boats, and she became well known to generations of tourists visiting the Highlands, for she ran until 1929. Larger vessels, paddle steamers, carried passengers along the Caledonian Canal. Curiously enough, despite the length of time (well over a century) during which Caledonian Canal steamers regularly carried passengers along Loch Ness, there are no reports of

29. Steamer Linnet *carried passengers along the Crinan Canal from 1866 to 1929*

monsters having been sighted from them . . . but then, perhaps the steamers frightened the beasts away below the surface!

Horse-drawn passenger boats (also called packet boats) saw their most exciting development on a short Scottish canal, the Glasgow, Paisley & Johnstone. Early passenger boats on canals appear to have been slow and ponderous things, but the committee of the Paisley Canal decided in the 1820s to experiment with various types of boat intended to travel, by the standards of the time, fast.

The boats that resulted were long, narrow and light in weight. Empty, they drew only $5\frac{1}{2}$ in. of water, and even with a full load of 90 passengers they drew less than 20 in. These boats were pulled by two horses at the gallop, the second ridden by a boy in a scarlet jacket; and they halved the time for the journey from Glasgow to Paisley.

This type of 'Scotch Boat' became famous and was introduced on other canals – among them the Lancaster and Kennet & Avon Canals in England and the Grand and Royal Canals in Ireland. Between Lancaster and Preston they averaged nine miles an hour over thirty miles, and since they had to stop for horses to be changed every four miles or so they must have gone faster over much of the distance. This does not sound much compared with speeds of land and air travel now: but on the other hand, it is worth pondering over, as one chugs along to-day's silted canals at the regulation four miles an hour.

However, even the swift passenger boats were unable to compete with railway trains; and parts of the Paisley Canal itself were converted into a railway in the 1880s.

30. Barge Sabrina W *at Selby on the Selby Canal has been converted into a Youth Hostel*

Pleasure trips on canals, though uncommon until recent years, are not a new idea. In 1828, for instance, the Union Canal Company not only had morning and evening passenger boats from Edinburgh to Glasgow, but also, in summer, an 'additional boat for the accommodation of pleasure parties'. Horse-drawn pleasure craft were introduced on the beautiful canal at Llangollen in 1884 and their successors continue to run. In many places, organisers of Sunday School treats and similar outings used to charter a canal boat or barge. An ordinary working vessel was used, cleaned out for the occasion, and off the excursionists would go; to judge from old photographs, there was often standing room only!

In 1951 John James started to run pleasure trips on the Regent's Canal in London. His boat *Jason* was a former trading narrow boat fitted with seats in the hold and an awning to protect passengers from rain. Similar boats have been introduced in many other places, usually motor driven but sometimes horse-drawn. 'Jason's Trip' became popular enough for a passenger-carrying butty to be added. Before this, the BWB Zoo Waterbus service was introduced; it started with converted narrow boats but now uses vessels built for the job. BWB also has on the Regent's Canal the *Lady Rose of Regents*, a wide-beam craft chartered out for functions and entertaining. On the Regent's Canal and elsewhere there are boats and barges fitted up as floating restaurants – some are permanently moored, others cruise while the customers eat; and at Selby on the Selby Canal is the barge *Sabrina W*, converted and permanently moored as a youth hostel.

In the early 1950s there were few if any boats built specifically for pleasure cruising on canals. People who wished to cruise on them had two main choices. The first was to use a motor cruiser designed for coastal or river use, which happened to have the right dimensions. For narrow canals this meant boats that were not only narrow but also short in proportion – although as interest in canals developed, boatbuilders have produced narrow-beam cruisers for them, in wood and glass fibre. For wide canals the choice was less restricted, and many large motor cruisers were and are to be found on wide canals such as those of the North of England, the Grand Canal and the Caledonian Canal. The Caledonian and Crinan Canals are also much used by yachts passing through them.

The other main choice on narrow canals was to use a converted narrow boat: a former trading boat with a long cabin built over the hold. I do not know who first had the idea of doing this, but there were certainly a few such boats in the 1930s and possibly

31. A narrow-boat style pleasure cruiser on the Grand Union Canal

32. A scene from the past is recreated by a pair of narrow boats carrying coal for retail sale

33. Closing a lock gate behind a pair of camping boats

earlier. They combine a hull shape and material intended for canal use with plenty of living space. With the decline in narrow boat carrying, many narrow boats have been converted; some are privately owned, others act (often in pairs) as floating, mobile hotels. Sometimes the owner retains the original boatmen's cabin complete with traditional decoration. Leeds & Liverpool short boats and Grand Canal barges have been converted likewise.

From converted narrow boats, canal pleasure craft continued to evolve. The trouble is that full-length boats are rather long for amateurs to steer and handle. So their owners started to shorten them, before building the new cabin, by removing part of the hull. Then, in the mid-1960s, as narrow boats suitable for conversion became fewer and fewer, construction started of new boats: with robust steel hulls, strong enough for constantly passing through locks, based upon the hulls of narrow boats, but between 30 ft and 60 ft long, and with cabins containing all modern amenities. A boat of this type is the most suitable for cruising on the English canals, and it looks right on a canal, too. It is a popular form of hire craft.

There is, however, yet another form of narrow boat which is popular for cruising. This is the camping boat: a former trading vessel, traditionally decorated and with few alterations (some of them revert to the retail coal traffic in winter). The cruising party eat and sleep under canvas in the hold. Several firms hire them, and they are deservedly popular with youth groups.

There is one other group of vessels to be found on all canals: the craft used for the essential task of maintenance. They can be divided into two classes. Boats in the first class are those that actually do something – such as weedcutters, pile-driving boats, icebreakers and dredgers (once hand-operated, then steam, but now I think all motor). Boats in the second class are those that carry things, such as maintenance materials – BWB seems to favour narrow boats for this task, even on wide canals – or mud, carried from dredger to disposal point by mud-hopper and attendant tug. The second class includes the humble but invaluable punts, with their little cabins, on which maintenance gangs base themselves while at work.

6 *Features of Canals*

'A very curious process', wrote Queen Victoria in her journal for 1847 after passing through the locks on the Crinan Canal. In full: 'the eleven locks we had to go through – (a very curious process, first passing several by rising, and then others by going down) – were tedious. . . .'

Evidently we were not amused; but I am inclined to think that we had missed the point!

Locks are the most prominent feature of canals. With locks at intervals a canal overcomes changes in ground level; it can climb out of one valley, cross a watershed and descend into the next, or carry barges from somewhere at sea level to somewhere several hundred feet higher.

Between locks are level lengths of canal called *pounds*. A lock itself is no more than a short section of canal which can be cut off from adjoining pounds by solid watertight gates. The pound at one end of the lock is at a higher level than the pound at the other. At both ends of the lock are *paddles* which are used to let water into or out of the lock chamber, to make its water-level equal to that of the pound above or below the lock. To prevent water washing away the sides of the lock, they are made of brickwork or other solid material.

On English canals, most locks are not worked by lock-keepers but by crews of boats. On the Caledonian and Crinan Canals they are worked by lock-keepers, and the Grand Canal has lock-keepers too, but since each looks after several locks it is usual for boat crews to assist, or to work the locks themselves when the lock-keeper is elsewhere.

So when a boat is to ascend a lock it must usually put part of the crew ashore first. After it has entered the lock the gates are shut behind it and paddles at the top end of the lock opened to let water in. When the water is high enough the top gates can be opened and the boat continues on its way . . . after picking up the crew.

Going down, the reverse takes place. Paddles at the bottom end of the lock are opened to let water out.

I have assumed that the crew of the ascending boat found the water in the lock at low level. If the lock had been full, they would have had to 'empty' it before the boat could enter. For descending boats, vice versa.

Now for the important details.

Lock gates have balance beams which project over the bank. To open or close the gate, push with the small of the back

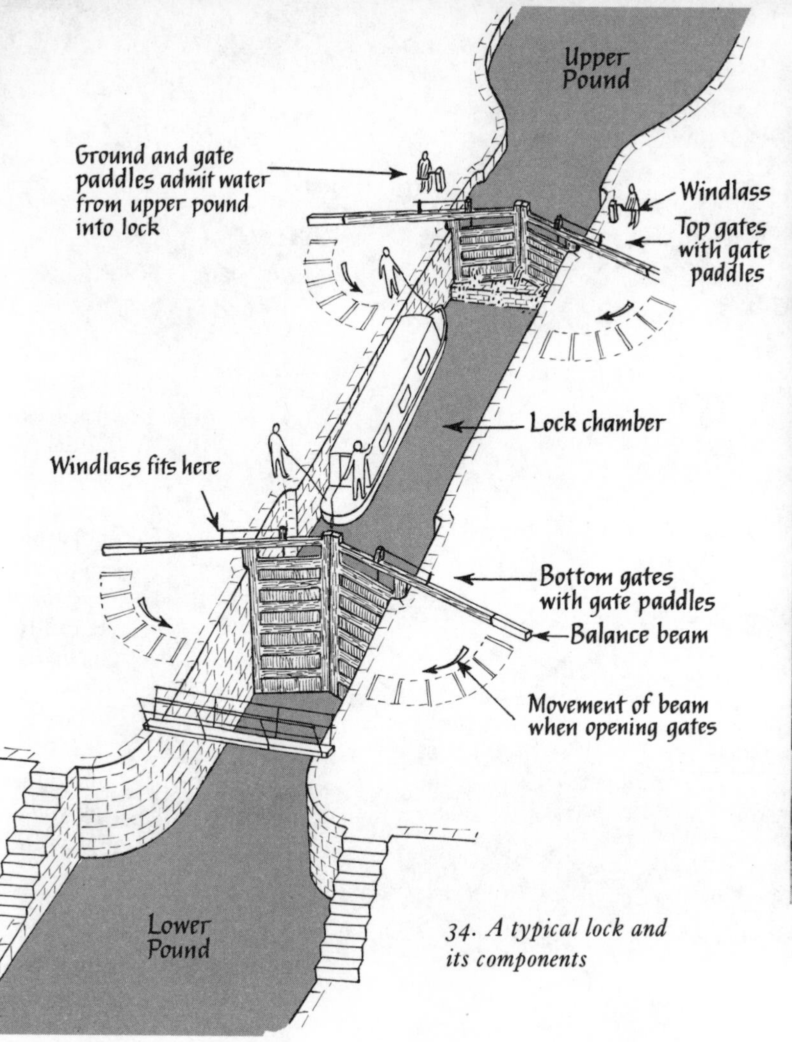

34. A typical lock and its components

35. Gate and ground paddles revealed on a canal drained for maintenance

36. Using a windlass to raise paddles

(preferably), or pull, against the beam near its outer end, steadily. An adult can manage this all right, but it sometimes needs two or more young people.

Paddles are the most mystifying parts of a lock, because many of them are hidden below ground or water or both. A paddle is just a sheet of timber which is raised to let water flow through a sluice or lowered to stop it. If the sluice is in a lock gate the paddle is called a gate paddle. If it is in an underground culvert which bypasses the gates the paddle is called a ground paddle. In each case it is connected to paddle gearing above, mounted either on the gate or on a post. To raise or lower the paddle a large L-shaped handle called a *windlass* has to be fitted on to a shaft and wound round and round. On the shaft is a gear wheel which raises or lowers a rack attached by a rod to the paddle below. Winding a windlass requires quite a bit of strength: many people under about 14 years old are not strong enough to raise paddles by themselves.

Two details are *really* important. Firstly, paddle gearing usually incorporates a safety catch. Make sure it is in position before raising the paddle, because if you let go of the windlass while doing so, and the catch is *not* in position, the weight of the paddle is almost certain to make it drop with a rush, with the heavy windlass whizzing round. The windlass is likely to fly off the shaft into the canal and be lost; alternatively it may hit you on the head, knock you out and send you to hospital, which rather spoils your holiday!

Secondly, before raising a paddle look at the *other* end of the lock to make sure its gates are closed and paddle racks down,

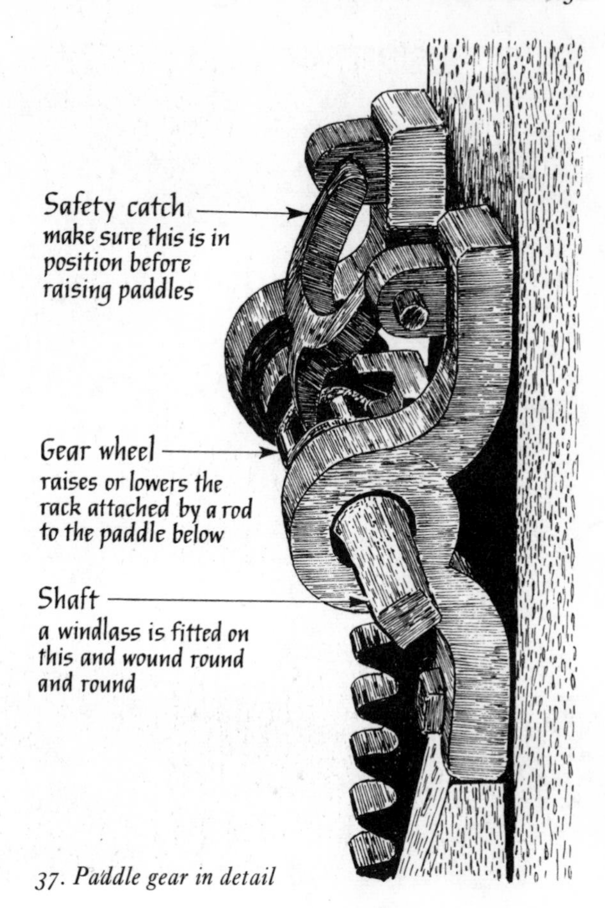

37. Paddle gear in detail

38. Inside the canal tunnel at Chirk on the canal to Llangollen

which shows that the paddles are closed too. Obviously, if paddles at both ends of a lock are opened, the water flows straight through. There is no danger of opening the *gates* at both ends of a lock at the same time because, unless the water level on each side of a lock gate is more or less equal, water pressure holds it shut.

When paddles are open, water flows fast through the sluices and then swirls about a lot. Many boats have to be held by ropes while ascending a lock to prevent their being swirled about too. When holding a rope, stand well back from the edge of the lock, for obvious reasons. When a boat is going down, a possible cause of trouble is the sill at the top end of the lock: this is a projection like a step (or, indeed, a window *sill*) which is in effect the end of the bed of the pound above the lock. A descending boat can get its stern or rudder caught on it as the water level drops, and so should keep away from the top gates. Whether a boat is going up or down, whoever is working the paddles should keep an eye on it to make sure it is all right.

After a boat has been through a lock it is best to leave the lock with all paddles closed and gates shut (unless there is a boat coming the other way). This reduces leakage.

These are the basic points of working locks: but this is not a complete guide. There are many variations in the details of lock equipment from canal to canal, and also in the terms used to describe them. Once you understand the principles, the best way to find out about working locks is to help someone experienced.

Where a canal has to rise steeply there are in places two locks, one after the other, with only three sets of gates. The middle gates are the top gates of the lower lock and, at the same time, the

39. Dundas Aqueduct carries the Kennet & Avon Canal across the River Avon near Bath

40, 41. Foxton inclined plane in 1905, with narrow boats waiting to enter

42. Foxton locks

bottom gates of the upper one. Where a canal rises still more steeply it may have several locks linked together in the same way, as a 'staircase'. At Banavie on the Caledonian Canal is a staircase with as many as eight lock chambers, and at Bingley on the Leeds & Liverpool is one with five. There is another at Foxton on the Leicester Section of the Grand Union which is really two staircases of five locks each, separated by a short pound in which boats can pass one another.

At the beginning of this century Foxton locks were bypassed by an inclined plane: a sloping railway up and down which boats were raised and lowered in big tanks of water. This took 12 minutes compared with about 1¼ hours to go up or down the locks. Unfortunately the inclined plane was too expensive to run and was abandoned after a few years, and the locks came back into use. Inclined planes of various sorts were used successfully on tub-boat canals but none now remains in use in the British Isles, though modern ones have been built for large canals on the Continent. However, at Anderton in Cheshire is the Anderton Lift, opened in 1875, which moves boats vertically in tanks of water between the Trent & Mersey Canal and the River Weaver more than 50 feet below.

Where a canal is ascending a valley alongside a river, its locks are often spaced out more or less regularly; elsewhere, canals were often built with flights of locks (several locks close together) and long pounds in between flights. The southern section of the Oxford Canal demonstrates these characteristics, and others too. Its summit pound is a good example of a contour canal: Brindley and some other early canal builders laid out their canals

43. Pleasure craft ascend wide locks at Soulbury, Grand Union Canal

44. Pontcysyllte aqueduct carries the canal to Llangollen high above the River Dee

along hillside contours, which enabled them to avoid extensive earthworks but made canals tortuous and long. The summit pound of the Oxford Canal is 11 miles long from Claydon Top Lock to Marston Doles, but the distance as the crow flies is no more than 5 miles. Later canal engineers built much straighter routes, using cuttings and embankments like the railway builders who followed them. The northern section of the Oxford Canal, from Braunston to Hawkesbury near Coventry, was straightened out like this in the 1830s.

Some parts of the old canal were left in use to serve wharves; where they join the newer canal, its towpath spans them by graceful bridges of cast iron, which are typical of their period. Bridges on the southern section are typical of their period, too. They are humped bridges made of brick or Cotswold stone, according to what was available, or lifting bridges, made of wood to save money. These have to be opened by boat crews, which means landing, running along the towpath and pulling down one of the bridge's balance beams to raise its deck – and then, preferably, sitting on the beam so that the deck does not come down on the boat as it passes beneath.

The balance beams are always on the side of the canal opposite to the towpath, which means you have to cross over the bridge before opening it and then cross back again afterwards. There are several other types of lifting bridge on canals, and swing bridges too, which swing horizontally, and most have the same characteristic. It seems stupid, until you realise there was a good reason: if bridges had been pivoted on the towpath side they would have been in the way of the towropes of horse-drawn boats, and each time a boat passed a bridge its towrope would have had to be detached.

There are reasons for most things about canals, though often they are not obvious. One of the fascinations of canal cruising is to work out what they are. Often they reveal something about the history of waterways. Along part of the canal now called the Grand Union there are mileposts lettered G.J.C.Co., because it was formerly the Grand Junction Canal. Canal bridges are usually numbered consecutively, but sometimes the numbers

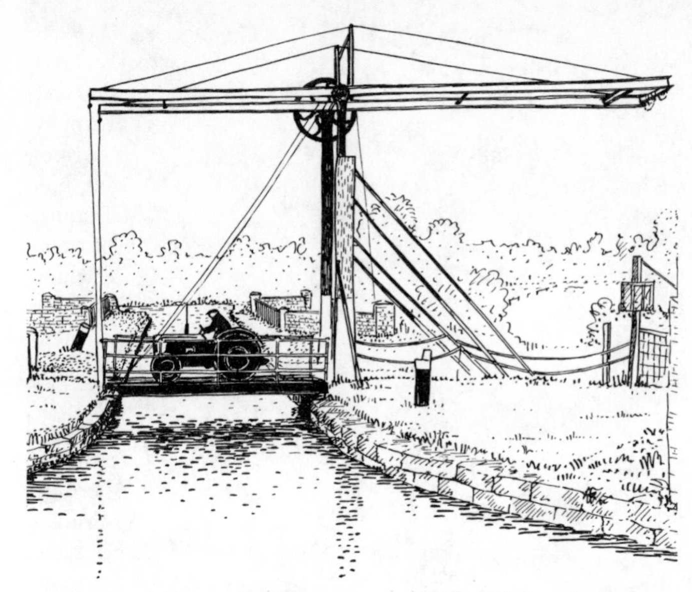

45. Lifting bridge at Monasterevin on the Grand Canal which crosses an aqueduct in the background

suddenly jump, say from 91 back to 1 (as at Braunston). This suggests that you have crossed the boundary between canals built by separate companies, though there may be no other obvious change.

Canal banks, particularly on the towpath side, are sometimes lined with steel or concrete piles. This is to prevent erosion caused by the wash of passing motor craft, which is often all too obvious on lengths which have not been piled. Silt washed away from the banks settles on the bed of the canal and makes it shallow so that it has to be dredged. The bed of the canal is made watertight by a lining of puddled clay.

Although canal pounds are level, they are not stagnant. There is always a slight flow of water. Summit pounds draw water from reservoirs and it flows along the pounds to make up for water used at locks, and lost by leakage and evaporation. Locks use a lot of water: even a narrow lock drains about 26,000 gallons from the pound above each time it is filled. Other sources of water are streams which flow into the canal and pumps which pump water up from a lower level, or back up a flight of locks. Nevertheless, during the dry summers of recent years, popular cruising canals have been short of water and so the hours during which locks could be used have been restricted.

Through hills too high for cuttings, canal builders bored tunnels; over rivers and roads, and valleys too deep for embankments, they built aqueducts. Going through a canal tunnel is a strange experience: to be in a boat, so much a thing of the open air, seems inconsistent with being underground in the dark. Canal tunnels range in length from a few yards to more than 1½ miles, and it takes more than half an hour to go through some of them. Early canal tunnels were built without towpaths, so people had to *leg* the boats through, by lying down head to head and pushing against the tunnel sides with their feet, while horses were led over the top. Later on, tugs were used.

The biggest and most impressive aqueduct is Pontcysyllte (pronounced, more or less, *Pont-kussuthlty*, with *u* as in *cut* and the last-syllable-but-one stressed). This is on the Llangollen branch of the Shropshire Union Canal. It was built by Telford

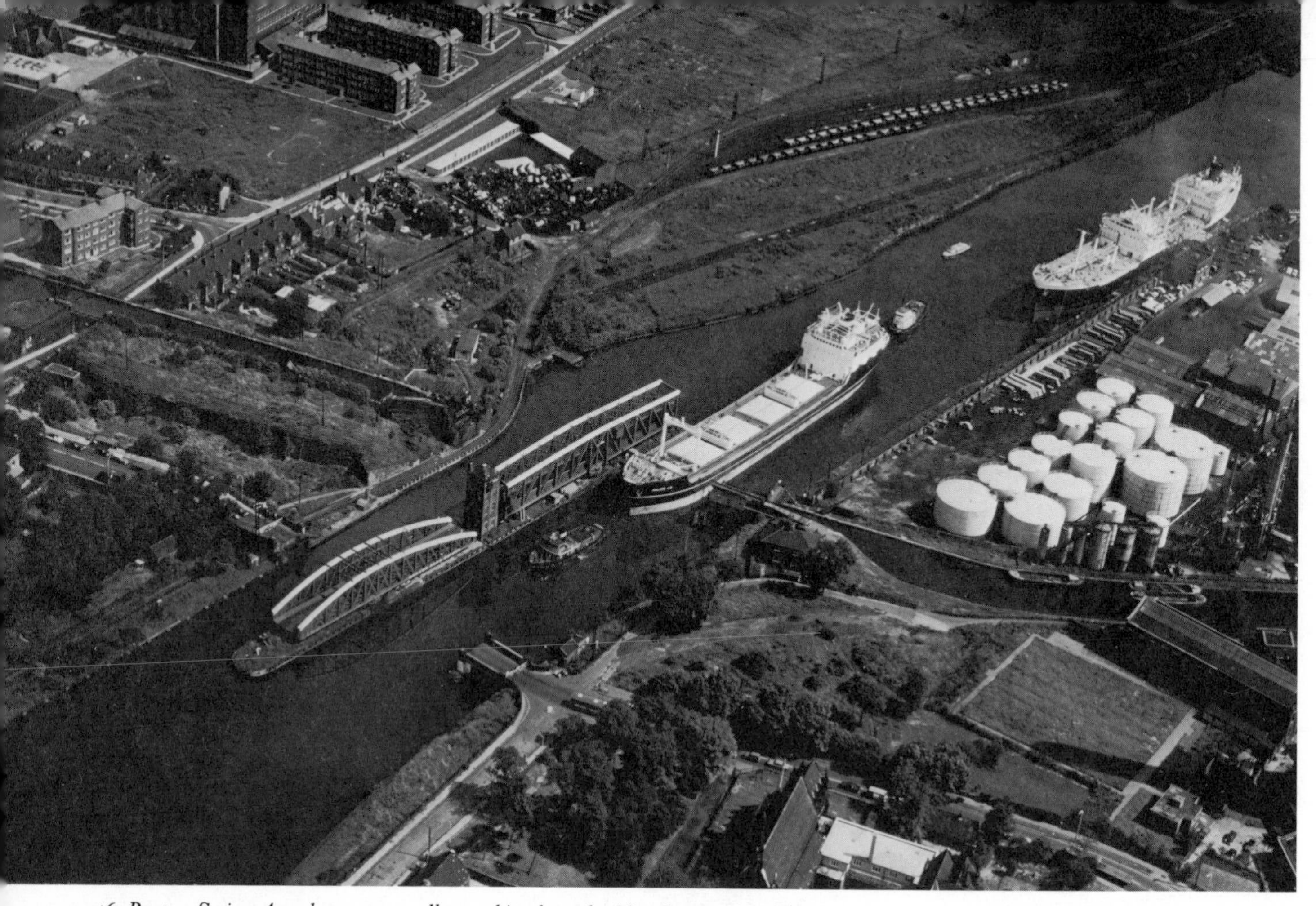

46. Barton Swing Aqueduct open to allow a ship along the Manchester Ship Canal

47. Warehouses and barges at Gloucester Docks, Gloucester & Sharpness Canal

between 1795 and 1805, has nineteen arches and carries the canal 121 ft above the River Dee. There is another high and imposing aqueduct at Chirk on the same canal, but second only to Pontcysyllte for grandeur are the Avon and Almond Aqueducts on the Union Canal in Scotland, which, since they are remote from the main English canal system, are less well-known than they should be. Barton Swing Aqueduct carries the Bridgewater Canal over the Manchester Ship Canal; like a swing bridge it swings open to let ships past. While it does so, gates like lock gates retain the water in the Bridgewater Canal, and on the aqueduct itself. It occupies the site of Brindley's original Barton Aqueduct. There are many other aqueducts, some of them fine pieces of engineering. Looked at along a canal, from boat or towpath, many appear to be no more than a narrowing of the channel. It is often worth stopping to examine them fully.

At intervals along canals are to be seen wharves and quays with cranes and warehouses. On commercial waterways barges load and unload cargo there. On cruising and remainder waterways they are now seldom used for their original purposes, but many are used by marinas and cruiser hire firms. Sometimes they are laid out round a canal basin, a large sheet of water connected to the main canal. In Ireland such basins are referred to as *harbours*, and in Scotland they are often called *Port* something-or-other! At intervals, too, come canal maintenance yards. With workshops, cranes, quays and maintenance craft, they remain true to their original purpose.

48. City centre: Gas Street Basin, Birmingham, in 1969

7 *How to Visit Canals*

Almost every mile of canal has a mile of towpath alongside, and the simplest and cheapest way of visiting canals is to walk along the towpaths. In fact the towpaths are not usually public rights of way but objection is seldom, if ever, made to walkers; and in some places – such as alongside the Regent's Canal in London – towpaths have been opened up as public footpaths. Elsewhere they are usually in good condition near locks but sometimes eroded or overgrown away from them.

The ideal way of visiting canals, though, is by boat. One has, of course, to pay for the privilege: either the boat must be licensed with the navigation authority or, on a few canals, tolls are paid for the use of locks. Probably the simplest and cheapest form of boat for canals is a canoe, and canoeists have in common with walkers the ability to explore canals which are not fully navigable, since their craft can be portaged past obstructions such as locks which no longer work.

For day excursions there are many passenger trip boats and restaurant boats. There does not seem to be a comprehensive guide to them published regularly, though one appears from time to time in the magazine *Waterways World*. Popular trip boats operate on canals from Little Venice (London), Bath, Stoke Bruerne, Llangollen, Dudley with its tunnel hewn through rock, Inverness on the Caledonian Canal and Robertstown on the Grand Canal. There are many others. A passenger boat equipped for handicapped children has been introduced on the Montgomery Canal near Welshpool, and there are occasional passenger trips on the Manchester Ship Canal, where pleasure craft are otherwise, understandably, unpopular with the canal company.

For longer cruises the main choices are either a family holiday – in your own family's motor cruiser if you are lucky, but otherwise, and no less fun, by hired cruiser or hotel boat – or an expedition by camping boat with school party or youth group such as Boy Scouts or Girl Guides.

Hired cruisers have accommodation for from two to ten people according to size of boat and are available throughout the English canals, and on the Caledonian and Grand Canals (the latter also being accessible to boats hired on the River Shannon). A catalogue such as that issued each year by Hoseasons, who take bookings for many firms which hire out cruisers, gives a good idea of what is available and where. More and more hire boats are fitted with cabin heaters for spring and autumn cruising, at which seasons they are cheaper than in summer. The actual hirer of a boat must usually be aged eighteen or over.

49. A motor cruiser explores the Barrow Line of the Grand Canal

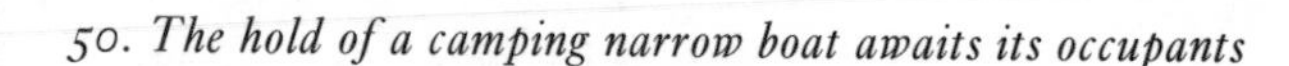

50. The hold of a camping narrow boat awaits its occupants

The most popular canals for cruising holidays are probably the Llangollen section of the Shropshire Union (because of its fine scenery and high aqueducts) and the southern section of the Oxford Canal. In fact these are now so popular that at the height of summer they are overcrowded. Other attractive and popular cruising canals include the Staffs & Worcs, the central part of the Trent & Mersey, and the Macclesfield; the Leeds & Liverpool is well spoken of, and the Ashby Canal is pleasant enough to deserve more use than it gets.

On a hired cruiser, the hiring party do everything. They form the crew: they run the boat, work the locks, cook the meals and so on. For many people such a holiday combines, in the right proportions, activity, idleness and a change from everyday life. For those who prefer less activity, hotel boats provide the change without the work, which is done by the paid crew – though even so they may, I suspect, welcome a helping hand with locks.

Camping boats can be hired with or without a boatman and there are many advantages in going on a pair complete with the boatman and his wife. The boatman is familiar with the route, points out features of interest, knows the best moorings and can even perhaps suggest a canalside field where a friendly farmer will permit a game of football when canalling, temporarily, palls.

Travelling on a camping pair is the nearest most people are likely to get to experiencing how pairs of working narrow boats used to be managed. To go through a flight of wide locks, for instance, motor boat and butty are 'breasted up', that is to say tied alongside one another; this means that they need only one steerer but have to move slowly. On pounds longer than about half a mile the boats are singled out, with the motor boat towing the butty – both have to be steered, but they go faster. There is much else to watch for, which the boatman can explain – not least, how to work locks properly.

The boatman and his wife generally live in their cabin, separate from the cruising party who live in the holds of the boats. A pair of camping boats can accommodate a party of up to twenty-four people; the holds contain beds, cooker, table, benches and so on. Protection from rain is provided by traditional side and top cloths into which transparent patches have been inserted to allow daylight through, while on fine days the cloths can be turned back to admit fresh air.

Another former trading vessel which has been converted to provide accommodation for youth groups, but which is otherwise

51. At Inverness on the Caledonian Canal: the Clyde Puffer Auld Reekie, *used by youth organisations, and the motor vessel* Scot II *which runs public cruises*

very different from a narrow boat, is the Clyde Puffer *Auld Reekie*. She has her original steam engine but the hold is converted into cabins; she is usually based in coastal water at Oban but her cruising range includes the Caledonian Canal.

For several years the Youth Hostels Association has organised adventure holidays aboard the horse-drawn converted narrow boat *Pamela*. The Duke of Edinburgh's Award Scheme, which is 'a challenge to young people' of fourteen and over, has a narrow boat cruiser *Lewis R. Jenkins*. This is usually kept on the Leeds & Liverpool Canal.

The clothes you need on a canalling holiday depend on the time of year, but although hot-weather clothes are needed in summer, it is a good thing to take some warm clothes too for cool evenings. These will certainly be needed for spring and autumn cruises. For wet days, a fully-waterproof anorak, or something similar, is essential. Preferably it should be accompanied by a sou'wester and waterproof overtrousers.

Shoes worn on boats should have rubber or rope soles so that you do not slip. Fashion shoes are seldom strong enough for canalling wear, and should be left at home; shoes with high heels of any sort are dangerous. Gumboots are useful in the mud and long wet grass of towpaths but cumbersome on board and inclined to weigh you down if you fall into the water.

A good torch is useful when going ashore after dark, and two essentials which organisers of cruising holidays sometimes forget are towels and matches. Remember also to take whatever you need for any other hobby which can be followed during a canal cruise – such as a camera, a flower press, a bird book.

8 *Boat Handling and Water Safety*

'What would you like me to do?' asked a visitor coming on board the author's canal cruiser recently. It struck me as a good opening: the speaker indicated that he understood that there was a skipper in charge of the boat, who would need the assistance of the others on board. This crew must follow his instructions and wait for them before doing things. All boats should have a skipper in charge, whether it is the owner, the head of the family or just whoever happens to be steering for the time being. Not that this gives the skipper the right to act as a little Hitler and roar orders at all and sundry, with the resentment this causes: he will get a better response if he *asks* people to do things, and says 'please' and 'thank you'.

When young people steer a canal cruiser or narrow boat there should be an adult nearby to advise, and take charge in emergency. So I give no detailed instructions here. But people on a canal boat do find themselves handling ropes and boathooks and so on, and to do so properly they should have some idea of the reasons behind the skipper's requests.

There are really three points to remember in which a boat differs from a land vehicle.

The first point is that it has no hand brake: it will not stay still for long unless moored, for wind and current quickly make it drift.

The second point is that, when boats pass other boats, the steerers normally keep to the right, not the left.

The third point is a bit more complicated, and is this: when a boat turns, it pivots about a point approximately one third of its length back from bow to stern (from front to back, that is). So when the bow turns one way the stern swings out to the other; and when the stern is moved to one side the bow points to the other. Not understanding this causes a lot of trouble to novices. For example, it is not uncommon to see, when a boat has come too close to the bank, someone standing at the stern who decides to be helpful and gives a shove off. All this does is to make the stern swing out and point the bow round *towards* the bank!

So if you think the boat ought to be pushed off, wait until the skipper asks, and then do so in the place and manner he suggests.

The skipper will usually need the assistance of the crew when leaving and approaching moorings, and when passing through locks and swing and lifting bridges. Locks and bridges I have already dealt with in chapter six.

Before leaving moorings the skipper will start the engine and let it run for a few minutes to warm up. Then he will ask for the mooring lines, at bow and stern, to be undone and brought on board, together with mooring spikes if they have been used. Because the edges of canals are often shallow, it is common to

52. On canals, boats keep to the right when passing

avoid damage to the propeller when setting off by pushing the stern of the boat out from the bank and then going astern (backwards) until the boat is in mid-canal, when it can go ahead. The mooring lines must be left ready for use: they may be needed in a hurry. This means they should be coiled, without knots or tangles, and put down somewhere handy. Other things should not be put on top of them, but neither must the ropes be left somewhere where they may be knocked or blown overboard, for a rope caught in the propeller sends a boat out of control.

Coming in to moorings is easiest against wind and current, in which event someone should take the bow line ashore when the skipper asks. If a strong wind from astern is unavoidable it is best to take the stern line ashore first, because to take the bow line first may result in the stern's swinging out across the canal. No one not actively involved in mooring the boat should attempt to leave it until it is securely moored – otherwise it may drift out from the bank just as he or she is attempting to step ashore . . . splash!

Which brings us to falling into the canal, and water safety generally. Sadly, people do drown in canals. In 1973, for instance, sixty-six people drowned in the canals and rivers of the British Waterways Board. You do not wish to add yourself to this statistic in future years, so be careful, and learn to swim. If you cannot swim, wear a buoyancy jacket when boating.

Although most cruising canals are not very deep, particularly at the edges, remember that locks are deep when they are full, and are difficult to get out of if you fall in when the water level is low. Avoid, particularly, falling in by an open paddle – there is very strong suction down through the sluice. If you fall in from a boat going along the canal, swim or wade to the towpath to be picked up: do not try to get back into the boat – you may become trapped underneath or be injured by the propeller.

When on a boat passing through locks or coming in to moor, do not dangle arms or legs over the side, lest they be crushed between boat and bank. When travelling on a cabin roof, watch out for low bridges, and branches of trees.

There is one other type of danger to beware of when canal cruising. Cooking and heating on many boats are by bottled gas. This is heavier than air, so if it leaks from, say, the cooker it sinks into the bilges of the boat, collects there and eventually, perhaps, explodes. Two precautions are important. When lighting the gas, strike a match before turning it on – rather than turning the gas on first to spill out while you fumble with a box of damp matches; and while the gas is burning, someone should always keep an eye on it to see that it is not blown out.

I have space only to give a broad outline of watermanship and water safety, coupled with some important details. Much more can be learned from water-based youth groups, such as sea scouts, who often have their bases beside canals, and the Pirate Club and similar clubs along the Regent's Canal in London. The Pirate Club, at Camden Lock, has rowing boats, canoes and a 30 ft pirate ship, complete with oars, sails and guns. Members join by turning up, but have to bring a paper signed by parents at, usually, their second visit. Swimmers aged eight to fourteen are admitted as pirates; younger children and non-swimmers as barge mice; and older members stay on as a senior group.

Activities of the Junior Section of the Calder Navigation Society, which is based on the Calder and Hebble Navigation in Yorkshire, include training members in boatmanship, lock procedure, rope handling and safety. Tuition has been given on board the Society's own boat *Doreen*, and the manner in which junior members handle her is said to put adults to shame!

9 *Canal Museums*

There are several museums wholly or partly devoted to canals and all of them are worth a visit. The most important are described below.

British Waterways Board's Waterways Museum at Stoke Bruerne, Northants, which is the goal for many school party visits, is housed in a fine old canal warehouse beside the Grand Union Canal. Popular exhibits are a full-size replica of a traditional narrow-boat cabin, complete with real decorations, furnishings and gleaming brassware; and an equally full-size replica of a boat horse, with genuine harness. Other exhibits include models, relics, paintings, engravings, old photographs and boat people's costumes. They bring 200 years of canal history to life.

A remarkable collection of old canal boats is being assembled for the Boat Museum at Ellesmere Port, Cheshire. This is being set up by the North West Museum of Inland Navigation Ltd jointly with the local authority and the first stage was opened to the public in 1976. Derelict canal basins and warehouses were being restored as a small traditional canal port and much of the work was being done by volunteers who included parties from local schools. The site combines old and new, for it is on the Shropshire Union Canal overlooking the shipping on the Manchester Ship Canal, and the boats include narrow boats of several types, a Leeds & Liverpool Canal horse boat, a tunnel tug and an icebreaker.

At Llangollen the canal wharf building now contains an excellent Canal Exhibition (not really a museum). The visitor, as he or she moves through a series of displays, is confronted by photographs, films, recordings, pictures, canal relics and working models which combine to tell the story of canals from the 1760s to the present day. At one point great mural paintings blend with real exhibits to give the impression of being on a busy commercial wharf.

Beside the Caldon Canal at Cheddleton, Staffs, is Cheddleton Flint Mill. It comprises a pair of water mills which formerly ground flint (a raw material for pottery); it has been restored, and is maintained in working order, voluntarily by the Cheddleton Flint Mill Industrial Heritage Trust. It is just the sort of small industry that canals were built to serve in the eighteenth century. Included in the restoration project was the mill's canal wharf and alongside this is moored the horse-drawn narrow boat *Vienna*, beautifully preserved in trading condition.

As mentioned earlier, the Blists Hill Open Air Museum (part of the Ironbridge Gorge Museum, Shropshire) includes a section

53. Passenger-carrying narrow boat Linda *comes alongside at the Waterways Museum, Stoke Bruerne*

of the Shropshire Canal, with a tub boat and an icebreaker. It also includes the site of the Hay inclined plane down which tub boats were lowered to a wharf on the River Severn, where their contents were transferred to barges. The inclined plane is being restored, and railway track re-laid on it.

The Exeter Canal is now little used commercially, but its basin at Exeter and warehouses alongside have become the home of the Exeter Maritime Museum. This is administered by the International Sailing Craft Association and is mainly devoted to a collection of sailing craft from many parts of the world, but it also includes a strange tub boat from the Bude Canal which was fitted with wheels to go up and down that canal's inclined planes.

54. Volunteers dig mud and rubbish from a canal basin which will form part of the Boat Museum, Ellesmere Port

55. Preserved narrow boat Vienna *at Cheddleton Flint Mill on the Caldon Canal*

10 *Restoring Disused Canals*

For more than a century canals were used less and less by boats. During that period some canals gradually became derelict and unusable. By the end of the Second World War the southern part of the Stratford-upon-Avon Canal, for instance, was impassable by anything larger than a canoe: wooden lock gates had decayed so as to be unusable, trees grew in the bed of the canal, and several short pounds were dry with cattle grazing on them.

Some canals were formally closed by Act of Parliament. The Forth & Clyde Canal was closed as recently as 1962, although boats were still using it, so that new roads which were planned could cross it by cheap low-level bridges without headroom for boats.

Eliminating disused or closed canals – that is, draining them, filling them in and demolishing locks and bridges so that they are not dangerous – is an extremely expensive business. So they tended to be left, neglected. Besides, many canals that were no longer used for transport had to be kept in existence for some other purpose. With the growth of interest in canals for amenity, such half-dead canals presented a challenge to many people who felt they should be fully restored, and that to restore and maintain them would be cheaper than to eliminate them. Backing words with action, people in increasing numbers started to work voluntarily to help restore canals.

The first successful waterway restoration project was the Lower Avon, a navigable river on which derelict locks and so on were restored voluntarily by the Lower Avon Navigation Trust between 1950 and 1962. A short piece of canal, the Wyken Arm, was restored voluntarily in 1959, but the big step forward for canals came with restoration of the southern Stratford Canal. This was transferred from the British Transport Commission to the National Trust in 1960, and after much voluntary work it was ceremonially reopened by HM the Queen Mother in 1964. Restoration of the Stratford Canal set an example which has been followed on many other canals, notably the Stourbridge, Peak Forest and Caldon Canals. In many places volunteers organised by the Waterway Recovery Group of the IWA have worked in conjunction with BWB staff on canal restoration schemes which have been aided with money provided by local authorities such as borough councils. In this way, many un-navigable remainder waterways have been restored for navigation. Campaigns continue for full restoration of other disused canals – such as the Montgomery, the Wey & Arun, the Forth & Clyde and, in Ireland, the Royal Canal.

Voluntary work on canal restoration by young people is

56. Caen Hill locks at Devizes, Kennet & Avon Canal. Taken when the canal was still navigable, this photograph shows how the side pounds were already overgrown

generally concentrated in organised groups such as school parties. The most notable organisation for matching the abilities of young people to the needs of canal restoration is the Junior Division of the Kennet & Avon Canal Trust Ltd.

The waterway called the Kennet & Avon Canal extends from Reading almost to Bristol. It is in fact part-canal, part-river, it is owned by British Waterways Board, and it has not been navigable throughout since 1950. Much of it deteriorated to such an extent that it could not be used. As a result of the efforts of the Trust, many of the locks and much of the waterway have been restored, and the work continues.

The main objects of the Junior Division are not only to involve young people in restoring the canal but also (and this is really more important) to help them develop character at the same time. So the tasks allotted become a form of adventure training which shows worthwhile results, and young volunteers do not become just a convenient form of cheap labour!

The division was founded in 1965 by Captain J. D. Mansfield-Robinson RN, who had been responsible for the introduction of adventure training on *Outward Bound* lines into the Navy for young sailors. Though it does have its own members, the division functions in addition as a centre to co-ordinate groups from many places who come to help restore the canal. They include scouts, sea cadets, and groups from schools and service training establishments.

The division's first achievement was to clear water weed and silt which were blocking the canal near Devizes, and so enable boats to reach that town and use the whole fifteen miles of the

57. Volunteers of the Kennet & Avon Canal Trust's Junior Division at work on one of the locks at Devizes

58. A disused section of the Kennet & Avon Canal near Bath awaits restoration

canal's *Long Pound*. Much of the work was done over several weekends in 1967 by *Operation Mudlark*. In a cutting which was particularly badly silted, young people with ropes dragged large scoops across the canal to fill them with mud; these were then taken by pontoons to a hoist which had been set up, and lifted up the side of the cutting to be tipped into a waiting vehicle.

The need to remove water weed which was choking the canal led the Junior Division to develop two floating self-propelled weed-removers, which it calls *Moonrakes*. I write *weed-removers*, rather than *weed-cutters*, because these machines have rakes or scoops which drag weed out by the roots and scoop it up from the surface. *Moonrake II* is the more advanced of the two machines, but although it incorporates a diesel engine and a complex electrical power-transmission system, it was assembled by pupils of four schools near the canal, with a little adult help, and can be operated by a crew of three or four young people after they have been trained.

The biggest obstacle to full restoration of the Kennet & Avon is the famous flight of twenty-nine wide locks at Caen Hill, Devizes. By 1969 much of the flight had been dry for many years and it appeared that some of the locks had become so derelict that they might prevent restoration altogether. But the Trust, after considering other possibilities such as some form of inclined plane, decided that it would be best, eventually, to restore them; and in the meantime it has given the Junior Division responsibility for preliminary restoration, up to the point of fitting new gates (which will be done professionally).

So the division's working parties cleared the area of jungly vegetation which was damaging the masonry of locks, and removed all rotten lock gates. Then they started to clear accumulated mud and rubbish from the lock chambers and the side pounds between the locks which are intended to act as small reservoirs. To remove spoil they have a narrow-gauge contractor's-type railway with small tipping wagons and a home-built diesel locomotive which can be – and is – driven by a young man aged twelve. Not only has the railway solved the problem of removing spoil, it also makes the work great fun!

Further Reading

For more information about canals, try the following books:

Hadfield, C., *British Canals*, David & Charles, Newton Abbot, 1950

Hadfield, C., *The Canal Age*, David & Charles, Newton Abbot, 1968

McKnight, H., *Canal and River Craft in Pictures*, David & Charles, Newton Abbot, 1969

Delany, V. T. H. and D. R., *The Canals of the South of Ireland*, David & Charles, Newton Abbot, 1966

Rolt, L. T. C., *The Inland Waterways of England*, Allen & Unwin, 1950

Rolt, L. T. C., *Narrow Boat*, Eyre & Spottiswoode, 1944

Most of these have been re-issued since the dates given.
Look also at the following periodicals:

Waterways News, British Waterways Board

Waterways World, Waterway Productions Ltd, Burton-on-Trent

Acknowledgments

Many people have helped me during preparation of this book, and I am particularly grateful to the following:

May Boyd, Bord Fáilte Éireann; Mrs Vera Bryant, The Royal Society for the Prevention of Accidents; Robert Copeland, Cheddleton Flint Mill Industrial Heritage Trust; Sheila Doeg, British Waterways Board; Mrs Ruth Heard, Inland Waterways Association of Ireland; A. J. Hirst, North West Museum of Inland Navigation Ltd; A. D. Hodge, Manchester Ship Canal Co.; R. J. Hutchings, Waterways Museum; R. W. D. Kirkham, Calder Navigation Society; Captain J. D. Mansfield-Robinson, CBE, RN, Kennet & Avon Canal Trust Ltd; Jenny McCafferty, British Waterways Board; Graham Palmer, Waterway Recovery Group; Mrs Ruth Rokicki and the staff of Union Canal Carriers Ltd; Viscount St Davids, Pirate Club; David B. Wain, Welsh Canal Holiday Craft Ltd.

Index